T0259705

SpringerBriefs in Applied Sciences and Technology

Forensic and Medical Bioinformatics

Series editors

Amit Kumar, Hyderabad, India
Allam Appa Rao, Hyderabad, India

More information about this series at http://www.springer.com/series/11910

Naresh Babu Muppalaneni
Maode Ma
Sasikumar Gurumoorthy

Soft Computing and Medical Bioinformatics

 Springer

Naresh Babu Muppalaneni
Department of Computer Science
  and Systems Engineering
Sree Vidyanikethan Engineering College
  (Autonomous)
Tirupati, Andhra Pradesh
India

Sasikumar Gurumoorthy
Department of Computer Science
  and Systems Engineering
Sree Vidyanikethan Engineering College
  (Autonomous)
Tirupati, Andhra Pradesh
India

Maode Ma
School of Electrical and Electronic
  Engineering
Nanyang Technological University
Singapore
Singapore

ISSN 2191-530X   ISSN 2191-5318   (electronic)
SpringerBriefs in Applied Sciences and Technology
ISSN 2196-8845   ISSN 2196-8853   (electronic)
SpringerBriefs in Forensic and Medical Bioinformatics
ISBN 978-981-13-0058-5   ISBN 978-981-13-0059-2   (eBook)
https://doi.org/10.1007/978-981-13-0059-2

Library of Congress Control Number: 2018939136

Printed on acid-free paper

This Springer imprint is published by the registered company Springer Nature Singapore Pte Ltd.
part of Springer Nature
The registered company address is: 152 Beach Road, #21-01/04 Gateway East, Singapore 189721,
Singapore

Contents

Chapter 1
A Novel Evolutionary Automatic Clustering Technique by Unifying Initial Seed Selection Algorithms into Teaching–Learning-Based Optimization

Ramachandra Rao Kurada and Karteeka Pavan Kanadam

Abstract This paper endeavors to embark upon one of the key inadequacies of simple k-means partitioning clustering algorithm; i.e., the number of clusters is precise and is yet to be initialized before the algorithmic execution, and the algorithms confinement of squared error function is toward local optima. This restriction may even affect to bestow on overlapping partitions in the given dataset; the anticipation from this premeditated work is to find automatically the optimal number of clusters, endorse the incurred clusters with cluster validity indices (CVIs), and eventually estimate the minimum consumption in percentage of error rate and CPU time when enforced over real-time datasets. This expectancy is down to earth with a unified approach entrenched into teaching–learning-based optimization (TLBO) by constituting an initial seed selection strategic algorithm at initialization step, thereby clusters configuration and affirmation in teacher and learner phases. Experimental results substantiate that this clustering framework efficaciously tackles the aforementioned limitations and capitulate promising outcomes.

Keywords Evolutionary Clustering · Automatic Clustering · Evolutionary Algorithms · Cluster Validity Index · k-means algorithm · Initial seed selection algorithms

R. R. Kurada (✉)
Department of CSE, Shri Vishnu Engineering College for Women, Bhimavaram, India
e-mail: ramachandrarao.kurada@gmail.com

K. P. Kanadam
Professor & HOD, Department of Computer Applications, RVR & JC College of Engineering, Guntur, India
e-mail: kanadamkarteeka@gmail.com

© The Author(s) 2019
N. B. Muppalaneni et al., *Soft Computing and Medical Bioinformatics*, SpringerBriefs in Forensic and Medical Bioinformatics, https://doi.org/10.1007/978-981-13-0059-2_1

1

1.1 Introduction

The natural evolutionary process is stimulated by evolutionary algorithms with a set of generic meta-heuristic optimization algorithms. These algorithms provide global optimum solutions with a complete set of solutions rather than a set of the single solution at the same time [1]. The most recently developed TLBO proposed by Rao et al. [2–8] is one among this category. This meta-heuristic algorithm differentiates itself by proving a single set of solutions at each independent run and lastly finds global optimal solutions over continuous spaces in solving both constrained and unconstrained real parameter optimization problems.

The principle behind clustering is to verify the intrinsic groups in a set of unlabeled data, where the objects in each group are indistinguishable under some decisive factor of similarity such as data elements within a cluster are more similar to each other than data elements in different clusters [9]. The CVIs are indices used to evaluate the ending outputs of any clustering algorithm and find the best partition that fits the underlying data [10–14]. In this work, at the cluster post processing phase, the CVIs like Rand index (RI) [10], advanced Rand index (ARI) [10], Hubert index (HI) [11], Silhouettes (SIL) [12], Davies and Bouldin (DB) [13], Chou (CS) [14] are used to evaluate the quality of clusters engendered and appraise the similarity of the pairs of data objects in diverged partitions.

In this paper, a novel evolutionary automatic clustering approach using TLBO algorithm is proposed. Broadly, this unified automatic clustering framework alienates TLBO into a clustering loom with an initialization step, teacher phase, and learner phase to produce optimal solutions. The inbuilt mechanism is underplayed with the single pass seed selection algorithm (SPSS) [18] and TLBO to form a unified automatic clustering framework. The SPSS algorithm is used in the initialization phase to automatically initialize the number of cluster centroids in the real-time datasets. In the teacher phase, initial seed selection algorithms SPSS is orchestrated to pioneer optimal partitions in the disruptive datasets. In learner phase, the clusters engendered are verified and re-validated using CVIs and other parameter indicators. Finally, the knowledge discovery tab holds the set of the final solution, which is collaborated with the optimal values in automatic clusters, minimum error rate, minimum usage of CPU time, CVIs, and other tacit parameter indicators.

1.2 Algorithm Constituents

Noble researchers Lloyd [15] and MacQueen [16] ascertained indispensable structure to a simple iterative clustering algorithm called simple k-means to detach a dataset into a valid number of partitions prescribed manually through user intuition. Due to the algorithm's super facial nature in its implementation, robustness, relative efficiency, and ease toward adaptability, k-means algorithms are commonly practiced by researchers across different disciplines in engineering and sciences. Despite these advantages, simple k-means do suffer from few limitations such as manual initialization of a number of cluster centers, confinement of squared error function toward local optima, randomly choosing of cluster centers, and sensitivity toward handling outliers and noisy data. So, in this paper, we attempted to conceptualize the provoked challenges of k-means with a novel automatic clustering algorithm AutoSpssTLBO to address these issues by unifying SPSS with TLBO.

k-means++ algorithm proposed by Arthur and Vassilvitskii [17], specifies a renovated procedure to initialize the cluster centers before proceedings to the iterations in standard k-means. This algorithm attains mixed results in distinct iterations by first choosing centroid and minimum probable distance to detach the centroid arbitrarily. The SPSS algorithm proposed by Kanadam et al. in 2011 [18] is one an initial seed selection method to initialize seed value for k-means and is also a modification method to k-means++. This method is robust to outliers and guarantees on the expected outcome with quality clusters. This algorithm is originated by altering the study made by Arthur and Vassilvitskii, 2007. SPSS initializes the first seed value and the minimum distance that separates the centroid for k-means++ based on the point to which is close to more number of other points in the dataset. Subsequently, SPSS emerged as a time-honored initial seed strategic algorithm to find an optimal number of automatic clusters automatically. Therefore in this work, this initial seed selection strategic algorithm SPSS is unified with TLBO to lionize best partitions in the given datasets. The SPSS initializes first seed and the minimum distance that separates the centroids based on highest density point, and which is close to more number of other points in the dataset.

The theory illustrated in TLBO is the replication of teaching and learning process accomplished in a traditional classroom. TLBO has two stages "teacher phase" or erudition from the teacher, and "learner phase" or exchange of information between learners. The supremacy of the teacher is deliberated in requisites of results or ranking of the learners. Moreover, learners also gain knowledge by interaction among themselves, which helps them in improving their results and ranks. Elitist TLBO algorithm is also established as successful evolutionary optimization algorithm to find global solutions with less computational effort and high reliability.

Teacher Phase It is the first part of the algorithm where learners learn through the teacher. During this phase, a teacher tries to increase the mean result of the classroom from any value A_1 to his or her level (F_A). But practically, it is not possible and a teacher can move the mean of the classroom A_1 to any other value A_2 which is better than A_1 depending on his or her capability. Considered A_j be the mean and F_i be the teacher at any iteration i. Now F_i will try to improve the existing mean A_j toward it, so the new mean is designated as A_{new}, and it gives the difference between the existing mean and new mean.

$$\text{Residue\_Mean}_i = r_i \left(A_{\text{new}} - \text{TF} A_j \right) \tag{1.1}$$

where TF is the teaching factor which decides the value of mean to be changed, and r_i is the random number in the range [0, 1]. Values of TF can either be 1 or 2 which is a heuristic step, and it is decided randomly with equal probability as:

$$TF = \text{round}[1 + \text{rand}(0, 1)\{2 - 1\}] \quad (1.2)$$

The teaching factor is generated randomly during the algorithm in the range of 1–2, in which 1 corresponds to no increase in the knowledge level and 2 corresponds to complete transfer of knowledge. The intermediate values indicate the amount of transfer level of knowledge. The transfer level of knowledge can be depending on the learner's capabilities. Based on this Residue_Mean, the existing solution is updated according to the following expression

$$L_{\text{new},i} = L_{\text{old},i} + \text{Residue\_Mean}_i \quad (1.3)$$

Learner Phase It is the second part of the algorithm where learners increase their knowledge by interaction among themselves. A learner interacts randomly with other learners for enhancing his or her knowledge. A learner learns new things if the other learner has more knowledge than him or her. Mathematically, the learning phenomenon of this phase is expressed below. At any iteration i, considering two different learners L_i and L_j where $i \neq j$.

$$L_{\text{new},i} = L_{\text{old},i} + r_i\left(L_i - L_j\right) \quad \text{if } f\left(L_i\right) < f\left(L_j\right) \quad (1.4)$$

$$L_{\text{new},i} = L_{\text{old},i} + r_i\left(L_j - L_i\right) \quad \text{if } f\left(L_j\right) < f\left(L_i\right) \quad (1.5)$$

1.3   Automatic Clustering Using TLBO Algorithm (AutoSPSSTLBO)

The proposed automatic clustering algorithm (AutoSpssTLBO) pursues a novel integrated automatic evolutionary framework by implanting SPSS an initial seed selection algorithm into TLBO algorithm. This theme concurrently optimizes CVIs as multiple objective functions and finds an optimal number of clusters automatically.

1.3.1   AutoSpssTLBO Algorithm

Step 1: Population P initialized randomly with n elements with m rows and d columns is

$$P_{i,j}(0) = P_j^{\min} + \text{rand}(1) * \left(P_j^{\max} - P_j^{\min}\right)$$

where $P_{i,j}$ is a population of learners and ith learner of the population P at current generation with g subjects is

$$(g) = \lfloor P_{i,1}(g), P_{i,2}(g), \ldots, P_{i,s}(g) \rfloor$$

Initialize each learner to contain $\max(k)$ number of selected cluster centers, while $\max(k)$ is chosen using **SPSS algorithm** within the activation thresholds in $[0, \infty]$. Choose a set C of k initial centers from a point-set $(x_1, x_2, \ldots, x_n)$, where k is a number of clusters and n is a number of data points:

Step 1.1: Calculate distance matrix Dist, where $\text{Dist}(i, j)$ represents the distance from i to j

Step 1.1.1: Find Sumv where $\text{Sumv}(i)$ is the sum of distances from ith point to all other points

Step 1.1.2: Find point i which is min (Sumv) and set index $= i$

Step 1.1.3: Add first to C as the first centroid

Step 1.2: For each point x_i, set $D(x_i)$ to be the distance between x_i and the nearest point in C

Step 1.3: Find y as the sum of distances of first $\frac{n}{k}$ nearest points from the index

Step 1.4: Find the unique integer i so that

$$D(x_1)^2 + D(x_2)^2 + \cdots + D(x_i)^2 \geq y > D(x_1)^2 + D(x_2)^2 + \cdots + D(x_{i-1})^2$$

Step 1.5: Add x_i to C

Step 1.6: Repeat step 1.2 through step 1.4 until k centers are found

Step 2: Find the active cluster centers with a value greater than 0.5 and keep such centers as a learner and keep them as elite solutions using Eq. (1.1).

Step 3: For $g = 1$ to $g_{\max}$ do

For each data vector P_g, calculate its distance from all active cluster centers using Euclidean distance.

Assign P_g to nearby clusters.

Modify duplicate solution via mutation on randomly selected dimensions of duplicate solutions before executing the next generation using Eqs. (1.2) and (1.3).

Step 4: Evaluate each learner quality and find teacher alias best learner using RI (default).

Step 5: Update the learners according to the TLBO algorithm using Eqs. (1.4) and (1.5).

Step 6: Repeat the procedure from step 2 to step 5 until a termination criterion is met. Report the final solution obtained by the global best learner (one yielding the highest value of the fitness function) at time $g = g_{\max}$.

1.4   Results and Discussion

This experiment evaluates the effectiveness of automatic clustering and initial seed selection algorithm, intermingled into TLBO after applying 50 independent runs over real-time [19] datasets. Table 1.1 consolidates the above metrics, and best entries are marked in boldface. The associating algorithms used in Table 1.1 are the three existing automatic clustering algorithms AutoTLBO [20], Automatic DE [21], Automatic GA

Table 1.1 Mean value of contrasting algorithms after completion of 50 independent runs over diverged datasets

Dataset (dimension)	Algorithm	No. of auto clusters	% of error rate	CPU time (s)	ARI	RI	SIL	HI	CS	DB
Iris (150 * 4)	AutoSpssTLBO	**3.01**	**08.17**	**167.34**	**0.8957**	**0.9737**	**0.0867**	1.0207	**0.8966**	**0.6230**
	AutoTLBO	3.02	11.00	195.23	0.8232	0.9733	0.0763	**0.9473**	0.7194	0.5178
	ACDE	3.06	12.67	397.86	0.8124	0.9613	0.0387	0.9226	0.7905	0.4373
	AutoGA	3.15	13.46	489.55	0.8144	0.9147	0.0219	0.8746	0.8358	0.6014
Wine (178 * 13)	AutoSpssTLBO	**3.13**	**30.60**	**234.96**	**0.6957**	0.7052	**0.3948**	**0.8957**	**0.7105**	**0.7550**
	AutoTLBO	3.20	41.22	260.12	0.6932	0.6842	0.3158	0.5683	0.3955	0.7693
	ACDE	3.26	32.02	537.62	0.4107	**0.7243**	0.2757	0.4485	0.3842	0.7326
	AutoGA	4.20	70.22	1149.4	0.5489	0.6779	0.1169	0.5631	0.3668	0.7136
Glass (214 * 9)	AutoSpssTLBO	**5.97**	**16.29**	**250.03**	0.6912	**0.7947**	**0.2953**	**0.6958**	**0.4895**	**1.2667**
	AutoTLBO	5.86	27.20	291.69	**0.7553**	0.7053	0.2947	0.4107	0.2991	1.3009
	ACDE	5.80	57.10	708.15	0.3266	0.7329	0.2671	0.4657	0.2745	1.2443
	AutoGA	5.40	68.78	963.25	0.4785	0.7234	0.2444	0.3999	0.2638	1.2847

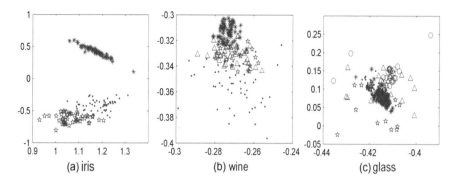

Fig. 1.1 a–c Clusters prompted by AutoSpssTLBO over wine real-time datasets

[22]. In case of real-time dataset iris, the proposed algorithm ISS-ATLBO produces an optimal number of automatic clusters as 3.01, minimal % of error rate as 8.17, and devours minimal CPU time as 167.34 s when compared with other algorithms in the regime. The mean value of ARI, RI, SIL, CS, and DB is optimal and is toward the near values to the upper boundaries of CVIs threshold. AutoTLBO algorithm makes its impact only in case of HI index with a mean value of 0.9473.

In case of wine dataset, the proposed technique claims an optimal number of clusters as 3.13, minimal error rate value as 20.31, and consumes low of CPU time as 234.96 s. The mean value of CVIs ARI, SIL, HI, CS and DB favors AutoSpssTLBO, and RI favors ACDE. In real-time dataset glass, the AutoSpssTLBO quotes an optimal number of clusters as 5.97, minimal percentage of error rate as 16.29, and low CPU time as 250.03 s. The dominance of AutoSpssTLBO sustains in all the comparing CVIs except ARI. The impression after testing the automatic clustering algorithms over real-time datasets is AutoSpssTLBO surmounts best results in the regime with respect to an optimal number of clusters, error rate, CPU time, and CVIs. Figure 1.1a–c characterizes the clusters prompted by AutoSpssTLBO over real-time datasets. These entire set of records justify the restrained stand of AutoSpssTLBO with a double standard that it can speedily adopt a comprehensive global convention in automatic clustering algorithms. The ending decree after comparing and contrasting the automatic clustering algorithms is the strategy of embedding initial seed selection in the a priori state is noteworthy. Hence, this anticipated technique AutoSpssTLBO is filled with a fulcrum of hopes that it can be applied to diverged domains of computer science.

1.5   Conclusion

This paper articulates a novel evolutionary automatic clustering algorithm by inter-mixing simple k-means partitioning technique and initial seed selection techniques into TLBO algorithm. The outcomes of this unified approach circumvents that the prospective algorithm has enough credibility in unraveling optimal partitions automatically in real-time datasets. Hence, the cessation annotations coagulated from this investigation are that AutoSpssTLBO algorithm has an immense periphery in portraying significant automatic clusters than its counterparts in the assortments assumed in this work. The subsequent purview of this work is to notice the fashionable consequences of the AutoSpssTLBO algorithm when expended over microarray datasets and other emerging domains in computer science. Also, the same perception of AutoSpssTLBO may be investigated again after blending in the flavor of Improved TLBO.

References

1. Bäck T, Fogel D, Michalewicz Z (1997) Handbook of evolutionary computation. IOP Publishing Ltd., Bristol
2. Rao RV, Savsani V, Vakharia DP (2011b) Teaching-learning-based optimization: an optimization method for continuous non-linear large scale problems. INS, 9211
3. Rao R, Patel V (2012) An elitist teaching-learning-based optimization algorithm for solving complex constrained optimization problems. Int J Ind Eng Comput 3:535–560
4. Rao R, Patel V (2013a) Comparative performance of an elitist teaching-learning-based optimization algorithm for solving unconstrained optimization problems. Int J Ind Eng Comput 4
5. Rao R, Patel V (2013b) An improved teaching-learning-based optimization algorithm for solving unconstrained optimization problems. SciIranica D 20(3):710–720
6. Rao R, Savsani VJ, Vakharia D (2011a) Teaching–learning-based optimization: a novel method for constrained mechanical design optimization problems. Comput Aided Des 43:303–315
7. Rao R, Savsani V, Balic J (2012b) Teaching learning based optimization algorithm for constrained and unconstrained real parameter optimization problems. Eng Optim 44 (12):1447–1462
8. Rao R, Savsani V, Vakharia D (2012a) Teaching–learning-based optimization: an optimization method for continuous non-linear large scale problems. Inf Sci 183:1–15
9. Jain K, Murty M, Flynn P (2002) Data clustering: a review. ACM Comput Surv 31(3)
10. Rand W (1971) Objective criteria for the evaluation of clustering methods. J Am Stat Assoc 66(336):846–850
11. Hubert L, Schultz J (1976) Quadratic assignment as a general data-analysis strategy. Br J Math Stat Psychol 29
12. Silhouettes R (1987) A graphical aid to the interpretation and validation of cluster analysis. J Comput Appl Math 20
13. Davies D, Bouldin D (1979) A cluster separation measure. IEEE Trans Pattern Anal Mach Intell 1(2):224–227
14. Chou C, Su M, Lai E (2004) A new cluster validity measure and its application to image compression. Pattern Anal Appl 7(2):205–220
15. Lloyd S (1982) Least squares quantization in PCM. IEEE Trans Inf Theor 28(2):129–137
16. MacQueen J (1967, June) Some methods for classification and analysis of multivariate observations. In: Proceedings of the fifth Berkeley symposium on mathematical statistics and probability, vol 1, No 14, pp 281–297, Chicago
17. Arthur D, Vassilvitskii S (2007) k-means++: the advantages of careful seeding. In: Eighteenth annual ACM-SIAM symposium on discrete algorithms. Society for Industrial and Applied Mathematics Philadelphia, PA, USA, pp 1027–1035
18. Kanadam KP, Allam AR, Dattatreya Rao AV, Sridhar GR (2011) Robust seed selection algorithm for k-means type algorithms. IJCSIT 3(5)

19. Blake C, Merz C (n.d.) UCI repository of machine learning databases. Retrieved 15 Oct 2014, from UCI repository of machine learning databases: http://www.ics.uci.edu/_mlearn/MLRepository.html
20. Kurada RR, Kanadam KP, Allam AR (2014) Automatic teaching–learning-based optimization: a novel clustering method for gene functional enrichments. Appears in "Computational Intelligence Techniques for Comparative Genomics", 2014. https://doi.org/10.1007/978-981-287-338-5
21. Das S, Abraham A, Konar A (2008) Automatic clustering using an improved differential evolution algorithm. IEEE Trans Syst, Man, Cybern—Part A: Syst Hum 38(1)
22. Liu Y, Wu X, Shen Y (2011) Automatic clustering using genetic algorithms. Appl Math Comput 218(4):1267–1279

Chapter 2
Construction of Breast Cancer-Based Protein–Protein Interaction Network Using Multiple Sources of Datasets

M. Mary Sujatha, K. Srinivas and R. Kiran Kumar

Abstract Biological data helps in disease analysis and drug discovery process. Study of protein–protein interactions plays a vital role to find the candidate proteins for causing chronicle diseases. Graphical representation of protein–protein interaction data gives more clues on protein functionality and pathways to research world. Many data sources are operating to provide such information upon request. However, each of them provides same information in different formats and levels. The word level denotes depth of interactions. This paper illustrates concatenation of different datasets related to target protein ERBB2 from prominent data sources. The resultant network highlights strong interactions of target protein after removing duplicate records with the help of computational tools and techniques, which makes work feasible and efficient.

Keywords Protein · PPI · STRING · BioGRID · Cytoscape · PID

2.1 Introduction

Biological data consists of different molecules among protein is one, and it interacts with other proteins to perform its functionalities. These interactions may be direct or indirect and can be visualized through a graph called protein–protein interaction network. Direct interactions are called as binary interactions, and indirect interactions refer to complex interactions [1]. An indirect interaction with target protein can be at any level of the graph. Inside protein–protein interaction (PPI) graph proteins act like vertices, and interactions among them represent edges. There are many databases [2, 3] which maintain PPI datasets based on publications and investigations made at the prediction of protein–protein interactions. Without having these data sources, it is very difficult to keep track all published interactions

M. M. Sujatha (✉) · R. K. Kumar
Department of Computer Science, Krishna University, Machilipatnam, Andhra Pradesh, India

K. Srinivas
Department of CSE, VR Siddhartha Engineering College, Vijayawada, Andhra Pradesh, India

© The Author(s) 2019
N. B. Muppalaneni et al., *Soft Computing and Medical Bioinformatics*,
SpringerBriefs in Forensic and Medical Bioinformatics,
https://doi.org/10.1007/978-981-13-0059-2_2

manually. Protein–protein interaction network representation varies from one data source to other, STRING [4]-like data source represents PPI graph with different levels [1] of interaction for a target protein to other proteins. Some other databases [5, 6] represent PPI graph inform of star topology where central node denotes target protein and other nodes of the graph represent source proteins. The volume of data provided by these repositories is a bit heavy, and there is much need to focus on highly interacting proteins, which helps in discovery of better drugs for a disease. An interaction between two proteins might be caused by satisfaction of different parameters. In recent days to extract high-dimensional interactions, data sources are utilizing computational techniques [7, 8] in predicting protein–protein interactions. Computational techniques are more feasible, cost-effective, and less time taking when compared with traditional laboratory-intensive techniques. Many data sources are also maintaining different filters in the prediction of protein–protein interactions such as score would be maintained for each interaction and highly scored interactions are taken into consideration while construction of a PPI network.

Apart from existing filtrations in the extraction of strongly interacting graph, combining multiple datasets together also provides high confidence interactions after eliminating duplicate nodes and interactions with the help of a graph analysis tool Cytoscape [9]. It is used to visualize biological graphs; further, graph manipulations and analysis can be done efficiently. The main purpose behind this paper is to construct a PPI network based on multiple datasets retrieved from different repositories and merged together to showcase strong interactions after eliminating duplicate entries. The same merge operation can be performed even by importing datasets directly from data sources too. However, this operation is limited with application registered databases only.

2.2 Materials and Methods Utilized

Breast cancer is one of the biological diseases commonly found among female worldwide. Based on earlier research and publications, ERBB2 is a target protein which causes breast cancer by its malfunction. P04626 is the protein identifier (PID) of ERBB2 recognized by UniProt data source [10].

2.2.1 Selection of Target Protein Associated Datasets

To study and analyze target protein functionality further ERBB2 interacting proteins retrieved from prominent databases namely STRING [3] and BioGRID [4]. STRING10.0 and BioGRID3.4 support many organism including Homo sapiens in retrieval of data. Interaction data can be downloaded in different formats. Disease-specific interaction data can be retrieved by quarrying database through a protein name or by group of proteins or by universally accepted protein identifier

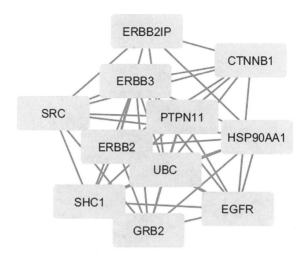

Fig. 2.1 STRING source network having 11 nodes and 46 interactions

(PID), each of which presented a different set of results in terms of format and level. After quarrying ERBB2 as target protein in STRING, it presented 11 nodes and 46 interactions among them (Fig. 2.1). The result is based on both physical and functional associations between proteins. Both computational techniques and laboratory inventions are used while predicting protein–protein interactions, whereas BioGRID showcased 172 nodes and 267 interactions among them (Fig. 2.2). The result is based on biological literature publications and by the contribution of other databases.

2.2.2 Data Concatenation

All the datasets retrieved from other databases need to be loaded into Cytoscape 3.5.0. It is also possible to interact with public databases directly from the application, if desired database is registered in it. Cytoscape 3.5 provides friendly navigator in uploading required datasets in any format. Network visualization and editing are possible for uploaded datasets. By concatenating retrieved datasets together, one common dataset is resulted after eliminating duplicate records. Figure 2.3 illustrates concatenation operation through venn diagram.

2.2.3 Reconstruction of Protein–Protein Interaction Network

Implementation of merge operation on considered two datasets produces the resultant network with common nodes and interactions. The output showcases a network with

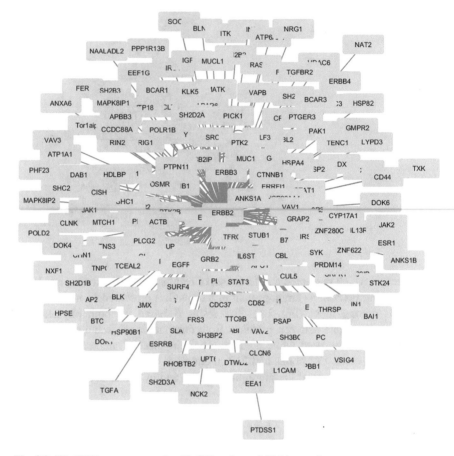

Fig. 2.2 BioGRID source network with 172 nodes and 216 interactions

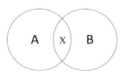

Fig. 2.3 Set A represents data source 1 (STRING), Set B represents data source 2 (BioGRID), and overlap of A and B represented by X denotes common data in both the sets

strong interactions. The same procedure is acceptable for more number of input datasets. The resultant network is reduced in its size having 10 nodes and 8 interactions among them (Fig. 2.4). It also produced outliers [11] representing common nodes and uncommon interactions. Protein PTPN11 acts as outlier as it appeared in both the input datasets without having common interacting protein.

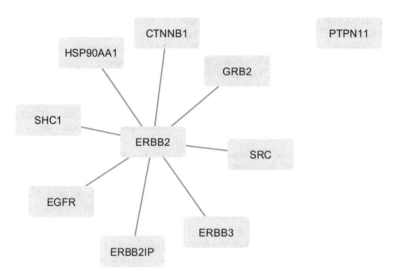

Fig. 2.4 Merged network with 10 nodes and 8 interactions

2.3   Conclusion

It is very much difficult and time-consuming to analyze each interaction of a large and complex network during drug discovery process. However, it is important to consider each interaction of a graph especially in analysis of biological diseases, and it is feasible to have small network with strong interactions. It can be achieved through graph manipulations. There are many factors that need to be considered during consolidation of datasets; for example, merge operation can be performed on similar type of data items. All data sources would not have similar type of data representation, especially source and target nodes that can be stored either with biological names or by universal identifiers-PIDs, though they represent same data. Data preprocessing is required to merge such kind of datasets, and it is left with a further study of research.

References

1. De Las Rivas J, Fontanillo C (2010) Protein–protein interactions essentials: key concepts to building and analyzing interactome networks. PLoS Comput Biol 6(6):e1000807. https://doi.org/10.1371/journal.pcbi.1000807
2. Zahiri J et al (2013) Computational prediction of protein-protein interaction networks: algorithms and resources. Curr Genomics 14:397–414
3. Mathivanan S et al (2006) An evaluation of human protein-protein interaction data in the public domain. BMC Bioinform 7:S19. https://doi.org/10.1186/1471-2105-7-S5-S19

4. Szklarczyk D et al (2014) STRING v10: protein–protein interaction networks, integrated over the tree of life. Nucleic Acids Res. https://doi.org/10.1093/nar/gku1003
5. Chatr-aryamontri A et al (2014) The BioGRID interaction database: 2015 update. Nucleic Acids Res 43:D470–D478. https://doi.org/10.1093/nar/gku1204
6. Orchard S et al (2013) The MIntAct project—IntAct as a common curation platform for 11 molecular interaction databases. Nucleic Acids Res 42:D358–D363. Database issue Published online 13 Nov 2013. https://doi.org/10.1093/nar/gkt1115
7. Rao VS, Srinivas K, Sujini GN, Kumar GN (2014) Protein protein interaction detection: methods and analysis. Int J Proteomics 2014, Article ID 147648 (Hindawi Publishing Corporation)
8. Mary Sujatha M, Srinivas K, Kiran Kumar R (2016) A review on computational methods used in construction of protein protein interaction network. IJEMR 6:71–77 (Vandana Publications)
9. Shannon P et al (2003) Cytoscape: a software environment for integrated models of biomolecular interaction networks. 13:2498–2504. ©2003 by Cold Spring Harbor Laboratory Press ISSN1088-9051/03
10. Apweiler R et al (2004) UniProt: the universal protein knowledgebase. Nucleic Acids Res 32 (Database issue):D115–D119. https://doi.org/10.1093/nar/gkh131
11. Rodrigues FA et al (2010) Identifying abnormal nodes in protein-protein interaction networks. São José dos Campos, SP, Brazil, July 26–30, 2010

Chapter 3
Design and Implementation of Intelligent System to Detect Malicious Facebook Posts Using Support Vector Machine (SVM)

Sasikumar Gurumurthy, C. Sushama, M. Ramu and K. Sai Nikhitha

Abstract Online social networks are the most widely used technology because of third-party apps. Huge people use social media to upload photographs and posts. Thus, hackers have decided to use social media applications for spreading malware and spam. So, we find that most of the applications in our dataset are malicious. Hence, we develop a tool called FRAppE—Facebook's Rigorous Application Evaluator. It is the first tool which is going to detect malicious applications. Malicious applications differ from benign applications in some features. So, in order to develop FRAppE, we first identify the set of features that help us distinguish malicious applications from benign applications. For example, malicious applications will mostly have the same name of benign applications. We consider mainly two parameters called app name similarity and posted link similarity. We use support vector machine (SVM) to classify between app name similarity and posted link similarity, and based on these results, we will detect the malicious post. Hence, we use FRAppE to reduce the use of malicious applications for accurate result. By including these parameters, the proposed framework produces better result prediction.

Keywords Online social networks · Third-party applications · Malicious applications

3.1 Introduction

Online social networks are most widely used because of its third-party applications. Users are attracting to these third-party applications because they are in the most entertaining way which helps us to communicate with friends, watch online shows, and can also play online games. So, hackers have decided to take advantage with this third-part apps and started to spread spam through these third-party applications.

S. Gurumurthy (✉) · C. Sushama · M. Ramu · K. S. Nikhitha
Sree Vidyanikethan Engineering College (Autonomous), Tirupati, India

© The Author(s) 2019   17
N. B. Muppalaneni et al., *Soft Computing and Medical Bioinformatics*,
SpringerBriefs in Forensic and Medical Bioinformatics,
https://doi.org/10.1007/978-981-13-0059-2_3

3.1.1 Face Book Apps

An interactive software application is developed to utilize the core technologies of the Face book platform to create an extensive social media framework for the app. Facebook apps integrate Facebook's news feed, notifications, various social channels, and other features to generate awareness and interest in the app by Facebook users.

3.1.2 Malicious

Malware, short for malicious software, is any software used to disrupt computer or mobile operations, gather sensitive information, gain access to private computer systems, or display unwanted advertising.

3.1.3 Online Social Network

The definition of online social networking encompasses networking for business, pleasure, and all points in between. Networks themselves have different purposes, and their online counterparts work in various ways. Loosely speaking, a social network allows people to communicate with friends and acquaintances both old and new.

3.1.4 Spam

Spam [2] is flooding the Internet with many copies of the same message, in an attempt to force the message on people who would not otherwise choose to receive it. Most spam is commercial advertising, often for dubious products, get-rich-quick schemes, or quasi-legal services. Spam costs the sender very little to send—most of the costs are paid for by the recipient or the carriers rather than by the sender.

3.2 Literature Survey

The author Goa found the objects for spam classification which are used for spam campaigns [3], instead of individual spam messages. The existing system can identify six features which are used to separate spam messages from benign messages. But, in the proposed system, the author developed an online spam filtering which is easily used and applicable at server. In this proposed system we can avoid frequent retraining and also there is no need for the presence of all spam in the training set. It has a set of new attributes which help to recognize spam accounts, and it secures from fraud in all ways by displaying a message called spam. The author Goa developed similar textual description collection method used for the cluster of wall posts. In the existing system, they used edit distance method to find the similarities. The author modified the existing system and proposed fingerprint method to identify the similarities. This fingerprint method is efficient for computation, and it mainly concentrates on two techniques. They are text similarity and posts of clusters which share URLs of same destination. The author K. Lee developed a social honeypot-based approach which helps to detect the social spam. This honeypot-based approach helps to identify the account of spammer, who are harmful for online social networks. This system works in two steps. In the first step, it deploys the social honeypots from online social communities, and in the second stage, it does statistical analysis. The first stage is used to identify the spam profiles in online social networks, and by using this output in the analysis phase, we will get the spam classifiers.

The author S. Lee developed an URL detection technique. We use this technique for Twitter. The name of this technique is WARNINGBIRD[5] which is durable, and it does not depend on the malicious page. This technique works on correlation which is useful for real-time classification and also has high accuracy. It can access all the tweets from timeline and categorize them based on the URLs. The proposed technique has huge correctness and highly accurate and low false rates. This technique focuses mainly on the assets to detect URLs which are malicious. Thus, the proposed technique based on URL detection for Twitter works well and gives better results. The author L. K. Saul proposed a technique called supervised learning technique which helps to detect malicious URLs automatically. In order to use the technique for URL detection, it uses lexical and host-based features. This technique also detects the accurate learning model with the help of mining and by analyzing the features which helps to detect the malicious URLs.

The author M. S. Rahman developed a Facebook application called MyPageKeeper. This application protects the user from malware and spam. It is a threefold approach which works in three stages. In the first stage, software is designed which helps to provide protection to users. In the second stage, it detects the spam e-mails and malicious Web applications. In the third stage, the analysis of the system is done. Thus, the author developed a real-time application which helps to detect spam e-mails and malicious Web applications.

The author G. Stringhini developed a technique which helps to detect the spammers [4] in online social networks. This technique also gathers the messages from online social networks. This technique is able to detect the spam in both small scale and large scale. This technique is able to detect the account of spammer even if there is no information about the honey profile of spammer. We can also use this technique for detecting spammer accounts on Twitter. It was said that many spammer accounts were identified and closed with the help of this technique.

The author N. Wang developed an authentication technique for third-party applications. This technique restricts the use of third-party applications and also limits the ability to publish the applications. This technique alerts the users when there is any leakage of data with the help of Facebook security settings. It also provides the list of malicious users to the users of Facebook, so that they can be aware of them. so, finally, this technique makes aware of the malicious users of the online social networks and also tells about the security issues.

3.3 Methodology

The main idea of proposed system is to find the malicious applications which are going to spread through third-party apps in the online social networks. To detect these malicious applications, we use a tool called FRAppE [1], Facebook's Rigorous Application Evaluator, which works based on support vector machine (SVM) classification technique. SVM is a binary classification technique and works based on two features, app name similarity and app link similarity. In the dataset, it contains all the information about malicious applications, and by using SVM classification technique, we will compare the app name similarity and app link similarity and detect the malicious applications. Previously, there are techniques to detect the malicious apps averagely but cannot detect the malicious apps particularly. This technique improves the accuracy and can particularly detect the malicious applications with low false positive rates. In the proposed system, we also collect all the feedback about the application from user side and use this information for the detection of malicious application. With the help of FRAppE tool, the user can also inform the newly detected malicious application by using report spam option. Hence, the tool FRAppE detects the malicious application and secures from spreading spam and malware through this malicious application (Fig. 3.1).

Algorithm procedure:

1. Request the user to add new application.
2. The request is then transferred to application server.
3. Then, the application tokens are transferred to the tool FRAppE.
4. FRAppE detects whether the tokens are malicious are non-malicious.
5. If the application tokens are not malicious, then it sends back to the user, or else if it is malicious, then it displays that the particular application is malicious.
6. Repeat the Step 1 process in order to add the new application.

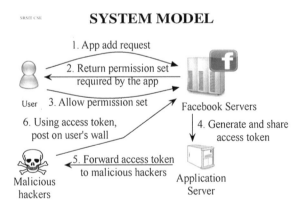

Fig. 3.1 System model to detect malicious applications

3.4   Modules and Discussion

Here, we have four different modules for the implementation of FRAppE, Facebook's Rigorous Application Evaluator, which is going to detect the malicious social media applications.

3.4.1   User

The user first logins to the system. In case if the user is a fresher, then he signs up into the system by agreeing all the terms and conditions. Then, the user will send a request to the system saying that he wants to add an application. The user should wait until he gets a reply from the system.

3.4.2   System Server

After getting the user request to add an application, the system server first verifies the user request and his/her registration details. After confirming the user, the system transfers his/her request to add application to application server. The request is transferred to the application server in the form of tokens which contains all the information about the user. The access to add an application will be provided only if the application server makes sure that the user is not a spammer and also only if the application is not malicious.

3.4.3   Application Server

Application server checks all the data about the application for which the user requested. It serves the data by checking the application name similarity and application link similarity with the applications present in the dataset. It also checks the application ID and the URL of the user and the application for which the user is requesting.

3.4.4   FRAppE

This tool works based on support vector machine (SVM) classification technique. It is a binary classification technique which considers application name similarity and application link similarity with the already present data in the dataset. So, with the help of this technique, we can find whether the application is malicious or not. If the application is malicious, then it does not provide access to use the application, and if the application is not malicious, then it provides access to use the application through application server. Here, we will discuss some measures through which the hackers try to evade the detection by FRAppE tool and also some tips on Facebook through which they can reduce the hackers on social media.

3.4.5 Similarity of Features

There is a possibility to evade the use of FRAppE in future with the help of the features that we are using for classification today. The features we are mainly using for classification are application name similarity and application link similarity. Along with application name similarity and application link similarity, we also say that malicious applications use the same name for benign applications. So the hackers can create a malicious application with the same name of benign application. Another feature is that malicious applications do not contain many posts. So, with the help of this information, hackers can develop an application containing many posts. There are also some features which will not be useful in the FRAppE and need to analyze the posts to detect whether an application is malicious or not. Finally, we need to conclude that the features used by FRAppE, such as URLs, permissions required, and client IDs for installation, are robust for the growth of hackers.

3.4.6 Tips to Facebook

There are two drawbacks through which the hacker can take advantage. The first drawback is malicious applications using a nonexisting client ID in the application installation which encourages the development of malicious applications. So, in order to avoid this, it should check that the client ID is similar to the application request client ID. Second, the Facebook should restrict the user to have arbitrary client IDs. One user should have only one client ID and should send request to access an application only through that client ID.

3.5 Implementation and Results

Here, we will consider the Facebook apps as datasets which contain content user reviews, apps, URLs, posts. With the help of this dataset, we are going to find whether the application is malicious are not. Here, we are estimating the results based on the contents of these datasets. We are going to prove that the proposed system will earn better results than the existing system.

3.5.1 Result

Table 3.1 represents accuracy between existing systems and proposed system on a different dataset. In our system, we use five datasets and perform the experiment. Our experimental result shows that proposed systems have high accuracy than the existing system because it added more features such as user reviews and description content checkup (Figs. 3.2 and 3.3).

Table 3.1 Accuracy comparison between existing and proposed systems

Datasets	Existing system (%)	Proposed system (%)
D1	82	87
D2	76	81
D3	81	88
D4	67	69
D5	85	90

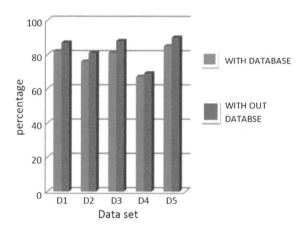

Fig. 3.2 Accuracy comparison between existing and proposed systems

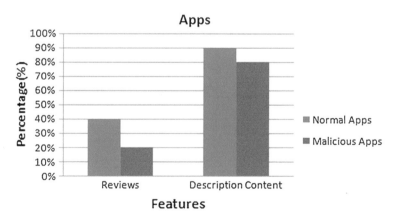

Fig. 3.3 Comparison between normal and malicious apps

Table 3.2 represents a comparison between normal apps and malicious apps based on user reviews and description content checkup. The malicious apps contain less features, and it has less users' reviews and description features. The normal apps have more features as compared to malicious apps.

Table 3.2 Comparison between normal and malicious apps

	Reviews (%)	Description content (%)
Normal apps	40	90
Malicious apps	20	80

3.6 Conclusion

Online social networks were most widely used because of its third-party applications. The third-party applications like Facebook, WhatsApp, Twitter are the most entertaining applications through which people can communicate with each other, play games through online, and can also spend their time by watching online shows. Because of these facilities, all the people are attracting to the third-party applications. So, hackers also decided to spread malware and spam through these third-party apps which are used by most of the people. We conclude that malicious applications have different features from benign applications. They are malicious applications that have the same name of the benign applications which is called as application name similarity, and also, malicious applications will not have many posts on the Facebook wall. So, with the help of these features, we can identify whether the application is malicious or not. With the help of these features, we created FRAppE which works on the classifier called support vector machine (SVM) which can give better results compared to the existing system. In the proposed system, we also add some features like application name similarity and application link similarity. With the help of these features, we develop a tool called FRAppE which will detect whether an application is malicious or not and gives better results compared to existing system.

3.7 Future Enhancement

In future, there is a chance to improve this technique by using clustering method. Clustering technique can identify malicious posts more effectively. We think that Facebook will get benefited by using our idea.

References

1. Rahman MS, Huang T-K, Madhyastha HV, Faloutsos M (2015) FRAppE: detecting malicious facebook applications. IEEE Trans Network 99
2. Gao H et al (2012) Towards online spam filtering in social networks. NDSS
3. Gao H et al (2010) Detecting and characterizing social spam campaigns. In: Proceedings of the 10th ACM SIGCOMM conference on internet measurement. ACM
4. Lee K, Caverlee J, Webb S (2010) Uncovering social spammers: social honeypots+ machine learning. SIGIR
5. Lee S, Kim J (2012) WarningBird: detecting suspicious URLs in Twitter stream. NDSS

Chapter 4
Classification of Abnormal Blood Vessels in Diabetic Retinopathy Using Neural Network

K. G. Suma and V. Saravana Kumar

Abstract Neovascularization is a serious sight-threatening complication of proliferative diabetic retinopathy (PDR) occurring in the diabetes mellitus persons, which causes progressive damage to the retina through the growth of new abnormal blood vessels. Preprocessing technique primarily extracts and normalizes the green plane of fundus image used to increase the level of contrast, the change in contrast level has been analyzed using Pair-wise Euclidean distance method. Normalized green plane image is subjected into the two-stage approach: detecting neovascularization region using compactness classifier and classifying neovascularization vessels using neural network. Compactness classifier with morphology-based operator and thresholding techniques are used to detect the neovascularization region. A function matrix box is added to categorize the neovascularization from normal blood vessels. Then, the feed-forward back-propagation neural network for the extracted features like number of segments, gray level, gradient, gradient variation, gray-level coefficient is attempted in Neovascularization region to get an indicative accuracy of classification. The proposed method is tested on images from online datasets and from two hospital eye clinic real-time images with varying quality and image resolution, achieves sensitivity and specificity of 80 and 100% respectively and with an accuracy of 90% gives encouraging abnormal blood vessels classification.

Keywords Diabetic retinopathy · Neovascularization · Normalization
Compactness classifier · Feature extraction · Neural network

K. G. Suma (✉) · V. S. Kumar
Department of Computer Science and Engineering,
Sree Vidyanikethan Engineering College, Tirupati, India
e-mail: sumarathee@gmail.com

V. S. Kumar
e-mail: bavya2002@gmail.com

© The Author(s) 2019   25
N. B. Muppalaneni et al., *Soft Computing and Medical Bioinformatics*,
SpringerBriefs in Forensic and Medical Bioinformatics,
https://doi.org/10.1007/978-981-13-0059-2_4

4.1   Introduction

Diabetic retinopathy (DR), the most common diabetic eye disease, occurs when blood vessels in the retina change which damages the microvascular system in the retina due to prolonged hyperglycemia, i.e., diabetes mellitus. India miserably is emerging as a world capital in diabetes by 2030 [1]. The occurrence at present has 40.9 million patients, the number that is expected to get higher to 69.9 million by 2025 [2]. Diabetes mellitus leads to affect eyes, kidneys, digestion, and skin. The Diabetic Retinopathy abnormalities do not show any early warning sign is majorly classified as non-proliferative diabetic retinopathy (NPDR) and proliferative diabetic retinopathy (PDR) [1].

NPDR abnormalities harms the retinal blood vessels arises to leak extra fluid and small amounts of blood into the eye [3], includes Microaneurysms, Hemorrhages, Exudates, Macular edema, Macular ischemia [3]. Many diabetic patients with mild NPDR do not get interrupted in their vision seriously. PDR abnormalities may cause severe vision loss than NPDR because it can affect both central and peripheral vision with the growth of new blood vessels at the back of the eye, the vessels swell and leak fluid or even close off completely [3]. PDR mainly occurs when normal blood vessels in the retina close and it halts the typical blood flow, responds the retina by growing new blood vessels is called neovascularization [3]. Thus, these new blood vessels are abnormal and do not direct the retina with proper blood flow. PDR includes vitreous hemorrhage, traction retinal detachment, neovascular glaucoma. Thus, it is stated that PDR condition which blocks the normal flow of blood toward the eye might lead to make drastic problem in human vision. Fundus photography was used to analyze the retinal part of an eye [4]. The proposed methodology uses a fundus camera image which snaps the interior surface of the eye used for monitoring (exudates, hemorrhages) and diagnoses the disease [4].

4.2   Literature Reviews

Neovascularization can be seen in iris, choroid, vein-ocular, corneal, and retinal within the part of eye. Upon that, the retinal neovascularization is the severe abnormality in PDR. The growth of these abnormal blood vessels has not been researched wide, but its severity is more which might lead to vitreous hemorrhage, microaneurysms, macular edema, etc. [5]. A number of studies related to DR with its abnormalities like exudates, microaneurysms, hemorrhages have been investigated in many papers in past decades.

Goatman et al. stated [6] that the neovascularization vessels classified according to their position: exists on or within optic disk are specified as New Vessels at the Disk (NVD) and otherwise they are specified as New Vessels Elsewhere (NVE). The paper [6] states that new vessels have a specific appearance: They are thin in caliber, more tortuous, and convoluted than normal vessels. Support vector machine classifier with 15 feature parameters was used to categorize the vessels as normal or abnormal. The work has been tested using 38 NVD images out of 71 DR images. The maximum accuracy is achieved at the sensitivity of 84.2% and specificity of 85.9%. Hassan et al. [7] detected the regions of both NVD and NVE using compactness classifier and morphological process. A function matrix box is used to mark the neovascularization vessel region from natural blood vessel. The method was tested on images from different databases which show the specificity and sensitivity results of 89.4 and 63.9%, respectively [7].

The main impact of this work is to classify the neovascularization where it has not yet been fully explored, and not many researches on its automatic detection were done. This two-stage approach which shows new vessels in both NVD and NVE is described and evaluated. Of all the abnormalities of DR, the growth of abnormal blood vessels carries the worst prognosis and the detection of this abnormality is most likely to add value to an automated screening system in future.

4.3 Data Description

4.3.1 Image Acquisition

The total fundus images with several abnormalities were collected from five online databases and two hospital eye clinics covering a populace of approximately 2400 peoples. A total of 970 images with DR were included in the dataset from three online databases along with two hospital eye clinics. In those 23 images from online database, 41 real-time images from hospital clinics are with confirmed neovascularization. IMAGERET–diaretdb0 (2006) is with 13 neovascularization images [8] and diaretdb1 with 2 Neovascularization images [9]. The online Messidor (2004) have only 8 images are with neovascularization [10]. Vasan Eye care (2012–2013) is with 35 neovascularization images. Bejan Singh Eye Hospital (2012–2013) is with 6 neovascularization mages. The data from three online database and two eye-clinic real-time images resembles a good practical situation, comparable and to evaluate the common act for analytical purposes.

4.3.2 Image Annotation

MESSIDOR [10] neovascularized images have been stated within the Excel file, and for diaretdb0 [8] database, ground truth documents indicate the neovascularization. Two experienced professional graders (Dr. M. Pratap, Ophthalmologist of Vasan Eye Care, Nagercoil, India, and Dr. Bejan Singh, Ophthalmologist, Bejan Singh Eye Hospital, Nagercoil, India) annotated the abnormal in freehand in online databases and for real-time images. The twisted, convoluted, busted growth pattern vessel segments are only considered abnormal by both graders and marked as abnormal.

4.4   Implementation

4.4.1   Preprocessing

Neovascularization is challenging toward the extraction of fundus image from its background, due to its busted growth pattern and the uneven illumination in the image [7] is shown in Fig. 4.1. Neovascularization blood vessels are twisted and thin in nature [7]. The green plane is extracted from the image used in the analysis since the green channel shows the highest contrast against the background is shown in Fig. 4.2a and the respective histogram is shown in Fig. 4.2b.

Figure 4.2a shows green plane looks clear than the red (eliminates the blood vessels presents in the optic disk) and blue (has very high contrast which submerges the blood vessels into its background) planes of the fundus image. Another progression has been done as the gray levels of green plane are normalized by a pre-specified mean and variance using Eq. (4.1).

Let $I(x, y)$ denote the gray value at pixel (x, y), M_i and V_i the estimated mean and variance of sector I, M_0 and V_0 are the desired mean and variance values, respectively. Let $N_i(x, y)$ denote the normalized gray-level value at pixel (x, y). For all the pixels in I, the normalized image is defined as (4.1). After normalization, the green plane is shown in Fig. 4.3a which gives the confidence level in finding the abnormal growth of blood vessels and the respective histogram is shown in Fig. 4.3b. The histogram Fig. 4.2b illustrates that the intensity values lie within 50–100th scale and the histogram Fig. 4.3b shows that the range of intensity values has been slightly adjusted toward the dark pixels and also shows a more balanced distribution between the dark and light pixels. The Histograms shows that the normalization process does not alter the shape of the original histogram plot, only the relative position of the values along the x axis is shifted; which simplify that the features presented in fundus image have not been changed, only the contrast gets increased by suppressing the background.

Fig. 4.1 Input image— Neovascularization fundus image

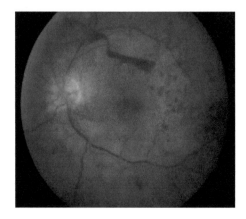

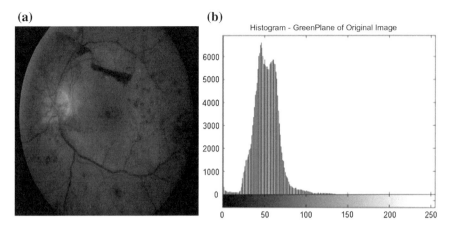

Fig. 4.2 a Green plane of input image. **b** Histogram of green plane of input image

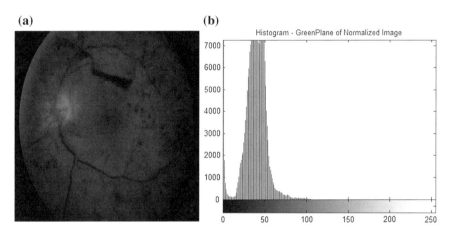

Fig. 4.3 a Normalized green plane image. **b** Histogram of normalized green plane image

$$
N_i(x, y) = \begin{cases} M_0 + \sqrt{\frac{V_0 \times (I(x,y) - M_i)^2)}{V_i}}, & \text{if}(x, y) > M_i \\ M_0 - \sqrt{\frac{V_0 \times (I(x,y) - M_i)^2)}{V_i}}, & \text{otherwise} \end{cases} \tag{4.1}
$$

4.4.2 Neovascularization Region Detection

Neovascularization is usually located at the connection of vasculature branches; it has no color and intensity differences with the dark lesions [7]. On the basis of

Fig. 4.4 Neovascularization
marked region

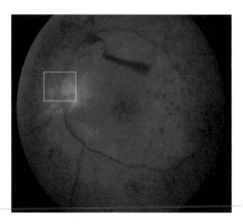

compactness-based classifier [7], the normal blood vessels are identified from abnormal blood vessels using Eq. (4.2).

$$\text{Compactness} = \frac{1}{4\pi \left[\frac{\text{perimeter}}{\text{area}} \right]} \qquad (4.2)$$

The fundus image of blood vessel might have bright lesions, dark lesions fragments, normal vessels, and also abnormal blood vessels. Multiple morphology dilation and erosion have been applied in the normalized green plane image in order to remove blood vessels other than neovascularization. Then the morphology thinning converts the object to lines of one pixel wide, morphology label used to calculate the number of blood vessels in the region, morphology area used to calculate the area of blood vessel covering the square window [7]. Neovascularization region detection method is intended using a square window with 100 × 100 size based on two assumptions:

(a) The 100 × 100 square window is passed through the abnormal blood vessels region, and it should contain at least four or more blood vessels. The numbers of blood vessels is calculated by separating the region of blood vessels.
(b) The 100 × 100 square window is passed through the abnormal blood vessels region, and it should contain more than 7% of the square window area. The area of blood vessels is calculated directly without separating the independent blood vessels region [7].

Based on the above conventions, the square window is used to detect the neovascularization region, and a box is strategic to spot the neovascularization by marking the region which is as shown in Fig. 4.4.

4.4.3   Feature Extraction

It is strategic to extract the features from the selected region in order to classify exact neovascularization from the detected region. The features [6] to detect the neovascularization in optic disk region are used here to categorize the neovascularization on NVD and NVE based on their shape, brightness, contrast and line density where gradient, gradient variation, gray level, number of segments; whereas other features based on tortuosity, ridge strength, vessel width, distance are not used for feature extraction because tortuosity and ridge related features are not needed for neovascularization region classification. However, those features help to classify neovascularization blood vessels on an individual basis. Finding at least one region with confirmed neovascularization is enough to help experts to treat patients. In this research work, features were extracted on the above idea; the statistical-based features were chosen for neovascularization feature extraction process. The five statistical-based features like number of segments, gray level, gradient, gradient variation, gray-level coefficient are used to illustrate neovascularization from each detected region. Strategic extraction of these features from the selected region is significant for accurate classification of neovascularization region.

4.4.4   Classification

The detection of neovascularization region sometimes detects non-neovascularization regions also as the abnormal blood vessels region as shown in Fig. 4.5. Thus, the classification of detection of such regions can be sorted out using feed-forward back-propagation neural network (FFBN). The paper [6] uses SVM classifier to classify the abnormal blood vessels that are present in optic disk alone got accuracy of 86.5% merely. FFBN is chosen for classification for its prompt

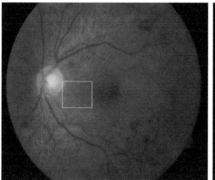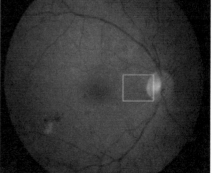

Fig. 4.5 Non-Neovascularization region marked as Neovascularization

training phase and good classification performance in which there is a feedback mechanism allowing the network to retrain it on data for adjustment of the internal weights. The input neurons hold the features data to be evaluated for each region to classify the neovascularization or not. Each value from the input layer is replicated and sent to all of the hidden neuron in the fully interconnected structure. The output layer transforms the hidden layer activations into the scale with less than 1.5 classifies as class 1 neovascularization region and greater than 1.5 classifies as class 2 non-neovascularization region. The neovascularization classification using FFBN gives the adequate accuracy performance of 90% to identify the abnormality evidently explains in the following section.

4.5   Results and Analysis

As described in Sect. 4.1, neovascularization is the severe one among the DR conditions. The abnormality is rare between people but it is serious one, when compared with other DR abnormalities neovascularization can be cured only through laser treatments. Fundus images are searched for this case and collected from three different public databases and two eye clinics. Thus, from the detailed selection process, 64 images are found with NVD and NVE. From that, 25 images are chosen for the implementation methodology. The low number of image selection for the implementation is to train and test the approach with correctly defined neovascularization declared images by ground truth and by professional graders as mentioned in Sect. 4.3.

4.5.1   Feature Selection

Initially, five features [6] like number of segments, mean vessel wall gradient, gray level, mean vessel width, mean ridge strength are taken into consideration and obtain only 70% of accuracy. The work is to classify in both NVD and NVE; thus, vessel-based features (mean vessel wall gradient, mean vessel width, mean ridge strength) vary in optic disk and pre-retinal regions provide low level of performance which is shown in Table 4.1. Hence, further three features like gradient, gradient variation, gray-level coefficient is taken into concern along with number of segments and gray level achieves the adequate performance of 90% is given in Table 4.1. The detailed performance of classification with selected features for each tested images is given in Table 4.2. The features which are the number of segments, gray level, gradient, gradient variation, gray-level coefficient of variation provide an adequate result to classify neovascularization in both NVD and NVE images.

Table 4.1 Performance comparison between features with vessel feature and selected features for classification

Features	TP	TN	FP	FN	Sensitivity	Specificity	Accuracy
Preceded vessel features (number of segments, mean vessel wall gradient, gray level, mean vessel width, mean ridge strength)	3	4	1	2	0.6	0.8	0.7
Proposed methodology vessel features (number of segments, gray level, gradient, gradient variation, gray level coefficient of variation)	4	5	0	1	0.8	1.0	0.9

4.5.2   Neural Network Classification

In the existing system [6], the neovascularization in optic disk was classified using special network—SVM classifier with 15 feature parameters got accuracy range of 84–86.5%. In this method, both NVD and NVE regions is classified using FFBN with five features got fine range sensitivity of 80% and specificity of 100% with 90% of accuracy is shown in Table 4.2 is calculated using Sensitivity, Specificity and Accuracy and with the same overhead comparison and the respective chart is shown in Graph 4.1.

The classification performance for FFN—Radial Basis function network (RB) got sensitivity of 60% and specificity of 80% with 70% of accuracy, and also verified with a special network Probabilistic neural network—Naïve Bayes classifier (NB) got sensitivity of 60% and specificity of 60% with 60% of accuracy are specifically tabularized in Table 4.3 with comparison chart is shown in Graph 4.2. Graph 4.3 shows the comparison chart for the accuracy levels between FFBN, RB, and Naïve Bayes which declares that FFBN gives adequate level in classification process.

Table 4.2 Classification performance using FFBN

Neovascularization detected region	Classification by FFBN
Diaretdb1–Image016	TP
Diaretdb1–Image027	TP
Diaretdb0–Image033	TP
Diaretdb0–Image034	TP
Diaretdb0–Image041	FN
Diaretdb0–Image002	TN
Diaretdb0–Image006	TN
Diaretdb0–Image007	TN
Diaretdb0–Image009	TN
Diaretdb0–Image010	TN

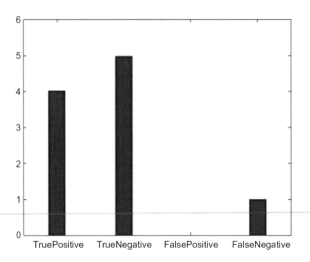

Graph 4.1 Classification chart using FFBN

Table 4.3 Classification performance comparison between FFBN, RB, and Naïve Bayes

Classification	TP	TN	FP	FN	Sensitivity	Specificity	Accuracy
Feed-forward back-propagation neural network (FFBN)	4	5	0	1	0.8	1.0	0.9
Radial basis network	3	4	1	2	0.6	0.8	0.7
Naïve Bayes classifier	3	3	2	2	0.6	0.6	0.6

Graph 4.2 Comparison chart for classification using FFBN, RB, and Naïve Bayes from neovascularization detected region

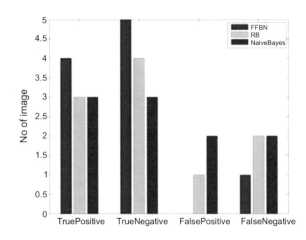

Graph 4.3 Comparison chart of accuracy achieved between FFBN, RB, and Naïve Bayes

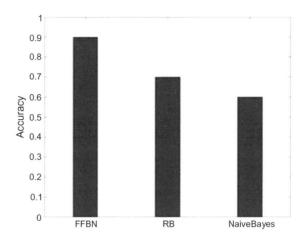

4.5.3 *Performance in Real-Time Images*

For online neovascularization images, the ground truth information is only mentioned in document statements and not having any region markings. Fluorescent angiography images are used to predict the presence of neovascularization in patients. Specialists treat the patients by injecting fluorescent dye toward the blood vessels which flows through the normal and abnormal vessels too that helps for laser treatments and not having any manual marking system to analyze the abnormality. Based on the statement of clinical specialist's free hand marking, the neovascularization region has been marked through Photoshop application, and then performances have been calculated by comparing with that markings.

The proposed methodology has been implemented in real-time images and achieves the same accuracy of 90% in classification elucidates in Table 4.4, the sample real-time image is shown in Fig. 4.6.

Table 4.4 Classification performance for real-time images using FFBN

Neovascularization detected region in real-time images	Classification by FFBN
Bejan Singh Eye Hospital Z59689	FN
Bejan Singh Eye Hospital Z70449	TP
Bejan Singh Eye Hospital Sankara Narayanan_Mr_05041973_G89605_83051	TN
Bejan Singh Eye Hospital Prema_Mrs_05061977_g95365_76773	TN
Bejan Singh Eye Hospital Sahaya Vinking_Mr_24041984_Y46073_93943	TN
Vasan Eye care 229247RajanMRAJM(90784)	TP
Vasan Eye care 233217ShowkathaliSSHOS(91193)	TP
Vasan Eye care 204418 204418SundaravalliRSUNR(68770)	TN
Vasan Eye care 8475SugumarRSUGR(98490)	TN
Vasan Eye care 8475SugumarRSUGR(98498)	TN
Sensitivity	0.75
Specificity	1.0
Accuracy	0.9

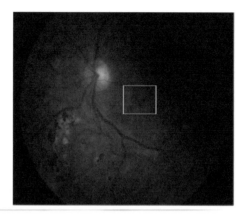

Fig. 4.6 Neovascularization marked (NVE) and classified in real-time image

4.6 Conclusions

The lack of research for neovascularization detection is because of the lower percentages of reported cases. The formation of online databases as mentioned in Sect. 4.3 was nearly before 10 decades. Currently, neovascularization creates a great focal topic because of its severity. Developing an automated system to find the abnormalities might reduce the cost of screening program. And doubted cases are referred to ophthalmologists rather than trusting on a screening system because of its inadequacy.

Yet, this approach would be helpful for the professional graders that it calculated a confidence measure of 90% in classification process. The developed work is stretchy and is able to handle with various image resolutions that are present in different clinical images. The difficulties in identifying Neovascularization by using proposed method, there may be a chance of not detecting neovascularization region due to the threshold value variations in region detection morphological process in images. And it is too challenging to define a common threshold value for all the databases because of the lack of neovascularization images. Hence, the method can be extended for the threshold range issue in future. This paper has validated an automated system which can able to classify the neovascularization vasculature in both NVD and NVE of DR images. It could form part of a system to reduce manual grading workload to analyze the serious DR abnormality to make patients to be attentive.

Acknowledgements The authors would like to thank Dr. M. Pratap, Ophthalmologist of Vasan Eye Care, Nagercoil, India, and Dr. Bejan Singh, Ophthalmologist, Bejan Singh Eye Hospital, Nagercoil, India, for their help in collecting the images.

References

1. Teena Thacker New Delhi (2009) India has largest number of Diabetes Patients: Report. The Indian Express, Wed Oct 21 2009, 03:34 hrs, Updated: Thu 20, Nov 2014, 11:15 Sun, 28 Apr 2013, 2:56 IST. Available online: http://archive.indianexpress.com/news/india-has-largest-number-of-diabetes-patients-report/531240/
2. Research letters (2012) A prescription survey in diabetes assessing metformin use in a tertiary care hospital in Eastern India. J Pharmacol Pharmacotherapeutics 3(3):273–275, July–Sept 2012. Available online: http://www.jpharmacol.com
3. Boyd K (2013) What is diabetic retinopathy. Eye Smart, Eye health information from the American Academy of Opthalmology, The Eye M.D Association, prohibited by US and international copyright law, 1 Sept 2013. Available online: http://www.geteyesmart.org/eyesmart/diseases/diabetic-retinopathy/
4. Cheriyan J, Menon HP, Dr. Narayanankutty KA (2012) 3D Reconstruction of human retina from fundus image—a survey. Int J Modern Eng Res (IJMER) 2(5):3089–3092. ISSN: 2249-6645, Sep–Oct 2012
5. Vislisel J, Oetting TA (2016) Diabetic retinopathy: from one medical student to another. EyeRounds.org, 1 Sept 2010. Available from: http://www.EyeRounds.org/tutorials/diabetic-retinopathy-med-students/
6. Goatman KA, Fleming AD, Philip S, Williams GJ, Olson JA, Sharp PF (2011) Detection of new vessels on the optic disc using retinal photographs. IEEE Trans Med Imaging 30(4):972–979
7. Hassan SSA, Bong DBL, Premsenthil M (2012) Detection of neovascularization in diabetic retinopathy. J Digit Imaging 25:436–444. Springer, Published online: 7 Sept 2011, Society for Imaging Informatics in Medicine 2011
8. DIARETDB0 database. Available online: http://www2.it.lut.fi/project/imageret/diaretdb0/
9. DIARETDB1 database. Available online: http://www2.it.lut.fi/project/imageret/diaretdb1/
10. MESSIDOR database. Available online: http://messidor.crihan.fr/

K. G. Suma received B.E. degree in Computer Science and Engineering in 2009, SMEC, Chennai, affiliated to Anna University: Chennai. She has received M.E. degree in Computer Science and Engineering in 2012, Muthayammal Engineering College, Rasipuram, affiliated to Anna University: Chennai. She is the university rank holder in PG. She received Ph.D. degree in Computer Science and Engineering from Anna University Chennai in 2017. Presently, she is working as an Assistant Professor in the Department of CSE at Sree Vidyanikethan Engineering College, Tirupati. Her area of research interest is medical image processing. She has published many papers in conference, national and international journals. She is a member of IEEE and ACM.

V. Saravana Kumar received his M.Sc. [Statistics] and M.Tech [Comp. Sci & IT] degrees from Manonmaniam Sundaranar University, Tirunelveli, in 1995 and 2008. At present, he is working as Faculty in Sree Vidyanikethan Engg. College, Tirupati. He had worked as Guest Faculty in center for BioInformatics, Pondicherry University in 2015. He had worked as Assistant Professor in CSE at Jayamatha Engineering College up to 2013, SCAD College of Engineering and Technology in 2009, and CMS College of Engineering in 2008. He has been published and presented good number of research and technical papers in international journals, international conferences, and national conferences. His area of interests are digital image processing, pattern recognition, algorithm, and data mining.

Chapter 5
Hybridizing Spectral Clustering with Shadow Clustering

N. Bala Krishna, S. Murali Krishna and C. Shoba Bindu

Abstract Clustering can be defined as the process in which partitioning of objects/ data points into a group takes place, such that each group consists of homogeneous type of data points and the groups must be disjoint of each other. Spectral clustering is the process in which it generally partitions the data points/objects into clusters such that the members of the cluster should be similar in nature. Shadow clustering is the technique in which it mainly depends on the binary representation of data and it is a systematic one because it follows a particular order for selection of index i such that whenever the main motivation is nothing but minimizing the required quality factor which measures the complexity level of irredundant final positive disjunctive normal form (PDNF), in such a situation the size of the anti-chain is AC and the number of literals is P. The machine vision or machine learning problems need an efficient mechanism for the effective performance of finding or analyzing images. For that, here a hybrid technique is implemented. The hybrid technique is the combination of both spectral and shadow clustering techniques. The proposed system is implemented as hybridization of shadow clustering and spectral clustering for effective performance of finding/analyzing the images in machine vision or machine learning problems.

Keywords Image segmentation · Clustering · Spectral clustering
Shadow clustering · k-means clustering

N. B. Krishna (✉)
Sree Vidyanikethan Engineering College, Tirupati, India
e-mail: balu1203@gmail.com

S. M. Krishna
SV College of Engineering, Tirupati, India

C. S. Bindu
JNTUA College of Engineering, Ananthpuramu, India

N. B. Krishna
JNTU College of Engineering Hyderabad, Hyderabad, India

N. B. Muppalaneni et al., *Soft Computing and Medical Bioinformatics*,
SpringerBriefs in Forensic and Medical Bioinformatics,
https://doi.org/10.1007/978-981-13-0059-2_5

5.1 Introduction

Clustering is a well-known data mining technique which is used for grouping the objects/data points of similar ones in to a group. Image segmentation is nothing but the partition of the image in to several regions which are of disjoint type, but those partitioned ones should be of homogeneous type. Image segmentation is a crucial step for some state-of-the-art algorithms mainly for understanding high-level image/ scene based on three reasons:

(i)   A region can have coherent support that has an assumption as it consists of only single label which can act as a priority for a lot of labeling tasks.

(ii) The coherent regions that obtained in the previous implementations contains a fine consistent feature extraction such that it can be used for the pooling of several feature responses over a particular region for incorporation of surrounding contextual information.

(iii) When comparing with pixels, the small number of better similar regions gradually decreases the cost of computation for the purpose of successive labeling task.

The process of image segmentation can be done on several kinds of algorithms such as clustering, histogram-based, region growing techniques. By considering several kinds of algorithms, one algorithm has been considered as significant one which is nothing but clustering-based algorithm. It generally pays attention to the researchers for the purpose of image segmentation. Clustering is a most well-known data mining technique in which the process of grouping the data elements exhibits similar behavior. Here, there are several clustering algorithms implemented, in that one of the traditional clustering algorithms is k-means clustering algorithm [1]. The k-means clustering algorithm considers only the similarity to k (number of desired clusters) centers. However, the k-means clustering algorithm consists of a limitation as it is very ineffective in the process of finding clusters. Here for that, the spectral clustering has been implemented in which it can be used in so many areas. For example, it can be used in information retrieval and computer vision.

Spectral clustering can be used as a process by treating the technique of data clustering as graph partitioning problem and lacking the chance of assumption in the form of data clusters. Spectral clustering is the process in which it generally partitions the data points/objects into clusters such that the members of the cluster should be similar in nature and clusters must be disjoint with each other; i.e., data points/objects are dissimilar with the data points/objects outside the cluster.

Shadow clustering is the technique in which it mainly depends on the binary representation of data. This (shadow clustering) technique was proposed by Muselli and Quarati [2]. Shadow clustering is the process which involves the reconstruction phase such that it can be treated as a problem for reconstructing the AND–OR expression of a specific monotone Boolean function which starts from a truth table subset.

5.2 Related Work

There are many algorithms that were proposed for the clustering techniques, which are as follows:

(i) improved spectral clustering of SAR image segmentation [3],
(ii) scalable spectral clustering for image segmentation [4], and
(iii) partition clustering techniques for multispectral image segmentation [5].

In those algorithms, the main concept involves image segmentation which was done by using clustering algorithms.

The analysis of clustering algorithms was also illustrated in (1) performance analysis of extended shadow clustering [6], (2) active spectral clustering via iterative uncertainty reduction [7], and (3) fast approximate spectral clustering [8].

5.3 Proposed Methodology

Spectral clustering can be used as a process by treating the technique of data clustering as graph partitioning problem and lacking the chance of assumption in the form of data clusters. Spectral clustering is the process in which it generally partitions the data points/objects into clusters such that the members of the cluster should be similar in nature and clusters must be disjoint with each other; i.e., data points/objects are dissimilar with the data points/objects outside the cluster. Spectral clustering produces results which are better than the general clustering algorithms like k-means, mixture models. Spectral clustering uses the affinity matrix which consists of eigenvalues and eigenvectors. Generally, the affinity matrix can be constructed based on raw data. The spectral clustering process can be done in two main steps:

1. Preprocessing: The preprocessing is the phase in which the construction of the graph and also similarity matrix which represents the dataset can take place.
2. Spectral representation: In this process, the graph (obtained from preprocessing step) has to be considered and by that the associated Laplacian matrix should be formed and after that the eigenvalues and eigenvectors of respective Laplacian matrix should be computed. After that, each and every point should be mapped to a representation of lower-dimensional phase, and that may be based on one or more eigenvector and clustering.

The classified data points have to be assigned to one or more classes based on new representation which is nothing but homogeneous type.

Shadow clustering is the technique in which it mainly depends on the binary representation of data. This (shadow clustering) technique was proposed by Muselli and Quarati [2]. Shadow clustering is the process which involves the reconstruction phase such that it can be treated as a problem for reconstructing the AND–OR expression of a specific monotone Boolean function which starts from a truth table subset. The procedure that was implemented by using SC (shadow clustering) is a systematic one, and the procedure follows a particular order for selection of index i such that whenever the main motivation is to decrease the preferred quality factor to measure the complexity level of irredundant final positive disjunctive normal form (PDNF), in such a situation the size of the anti-chain is AC and the number of literals is P. There consists of two main important quantities, and those quantities have to be minimized for making the desired resulting positive Boolean function [2]. For successful implementation of the above goal, this (shadow clustering) technique adopts two possible criteria as follows:

1. First one is nothing but maximum covering shadow clustering (MSC) which is for increasing the number of points in s and t and those can be covered by each and every bottom point which was generated.
2. Second one is named as deepest shadow clustering (DSC) which generally reduces the degree of the produced prime implicant.

Here, there is one thing to notice that the main motivation of MSC plays a key role when the algorithm for partially defined positive Boolean function (PDPBF) reconstruction is used particularly in machine learning problem solution [9]. By applying the traditional clustering algorithm to machine learning problems or to machine vision, the criteria of effectiveness is very important since the finding of the result can be used for analyzation of several images. For achieving more effectiveness, a hybrid algorithm is implemented, and the hybrid algorithm is nothing but a combination of both shadow clustering and spectral clustering.

The hybrid algorithm is implemented in the way such that the spectral clustering and shadow clustering algorithms are effectively combined to perform effective image segmentation in machine learning and machine vision problems.

The algorithm consists of several steps, which are as follows:

The inputs of the algorithm are: n number of r representative points of an antibody should be consider, K no of clusters have to be consider, $q1, \ldots, qn$ be the sampling points of a dataset.

As the first step of the algorithm, the extrapolation matrix E has to be calculated based on considering the representative points r by Formula

$$E = K(qi, r)|cj| \tag{5.1}$$

where $|cj|$ represents cluster size with respective jth representative point and $i = 1,$ ..., m and $j = 1, ..., n$.

The quantization matrix S has to be calculated based on considering the representative points r by Formula

$$S = K(ra, rb)|cb| \tag{5.2}$$

where $|cd|$ represents cluster size with respective dth representative point and $a, b = 1, ..., n$.

The diagonal matrices of S and E are calculated as follows:

$$DS = \Sigma m \quad 1p(re)K(ra, rb) \tag{5.3}$$

$$DE = \Sigma n \quad 1p(re)K(qi, rj) \tag{5.4}$$

where $a = 1, ..., n$ and $i = 1, ..., m$ and p(re) and p(rf) are denoted as elements of diagonal matrix of quantum antibody population.

After that eigen decompositions of diagonal matrix DS have to be calculated, and those eigen decomposition values will extrapolate in to the matrix N, then normalize the matrix N and locate the largest eigenvector based on Kth cluster for construction of matrix F and the eigenvalue ranking takes place as mentioned in Fig. 5.3. There after every row of matrix F have to be considered and group them in to K clusters by using k-means algorithm and assign those clusters to T subsets. Assign the anti-chain AC to $\varnothing$ and assign T subsets to X.

If the collection of subsets X is not empty, then append T, G subsets to the anti-chain which is obtained by the Boolean lattice of elements of x. Then, delete all the elements from X which covers a bottom point in the anti-chain. Later the subset of anti-chain have to be find-out which is having minimal complexity and then elements of X must belongs to T and covers elements of subset of anti-chain. Then, by using subset anti-chain, the non-duplicate PDNF is constructed and the graph will be generated as shown in Fig. 5.4.

5.3.1 Hybrid Clustering Architecture

See Fig. 5.1.

5.3.2 Algorithm for Hybrid Clustering

Inputs:
Number of quantum antibody representative points (r) is n, and no of clusters K.

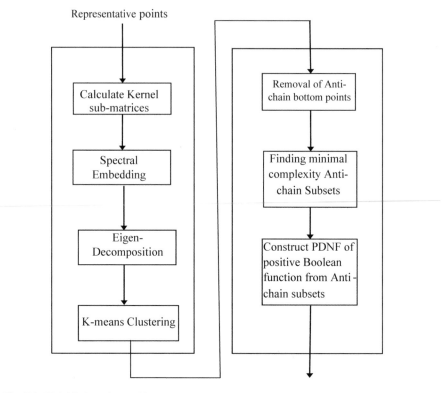

Fig. 5.1 Hybrid clustering architecture

Step 1:   Calculate the extrapolation matrix E and quantization submatrix S which are the basis of representative points r.

Step 2:   Calculate the diagonal matrices DS and DE.

Step 3:   Compute eigendecomposition of DS-1/2 S DS-1/2 $v = \lambda v$, extrapolate $N = $ DE-1/2 E DS-1/2 v $\lambda - 1$, normalize the matrix N, and locate the largest K eigenvector and store those results into matrix F.

Step 4:   Considering every row of F as respective point, group those points into K clusters by using k-means algorithm and assign those clusters to T subset.

Step 5:   Assign AC $= \Phi$ and $X = T$.

Step 6:   While X is not empty, do
           Select an $x \in X$.

Append to AC several bottom points for (T, G), obtained by descending from x the diagram of $\{0,1\}n$.

Delete all the elements from X which covers the bottom point in AC.

Fig. 5.2 Segmentation results of several clusters as histogram of self-tuning clusters

Fig. 5.3 Segmentation results of several clusters as eigenvalue ranking

Fig. 5.4 Average number of implicants that was generated in the process of reconstructing the Boolean function

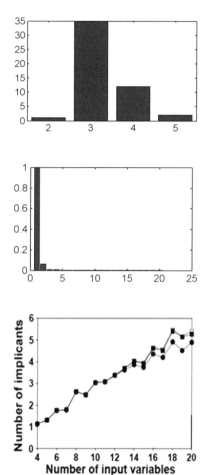

Step 7: Find out the subset AC* of AC such that having minimal complexity for every $x \in T$ covers some $a \in$ AC*.

Step 8: Construct from AC* the non-duplicate PDNF of the anonymous positive Boolean function p.

5.4 Conclusion

The effective clustering method of hybridization was implemented for the efficient performance of finding and analyzation of images in machine learning or machine vision. For that, an effective mechanism was implemented which is nothing but hybrid clustering. And the results produced were effective as shown in Figs. 5.2, 5.3, and 5.4. And the results were produced by an efficient proposed architecture.

References

1. Matthias B, Juri S (2009) Spectral clustering. Data Warehousing and Data Mining Free University of Bozen-Bolzano, 23 Jan 2009
2. Muselli M, Quarati A, (2003) Shadow clustering: a method for Monotone Boolean function synthesis. Rapporto interno IEIIT/GE/2/03—Agosto, pp 1–22
3. Gui Y, Zhang X, Shang Y (2012) SAR image segmentation using MSER and improved spectral clustering. J Adv Signal Process 2–9
4. Tung F, Wong A, Clausi DA (2010) Enabling scalable spectral clustering for image segmentation. In: Enabling scalable spectral clustering for image segmentation, vol 43, pp 4069–4076
5. Nuzillard D, Lazar C (2007) Partitional clustering techniques for multi-spectral image segmentation. J Comput 2(10)
6. Senguttuvan A, Krishna PD, Rao KV (2012) Performance analysis of extended shadow clustering techniques and binary data sets using K-means clustering. Int J Adv Res Comput Sci Softw Eng 2(8):51–62
7. Wauthier FL, Jojic N, Jordan MI (2012) Active spectral clustering via iterative uncertainty reduction. In: International conference on Knowledge discovery and data mining, vol 1, no 6, pp 1339–1347
8. Yan D, Huang L, Jordan MI (2009) Fast approximate spectral clustering. In: Proceedings of the 15th ACM conference on knowledge discovery and data mining (SIGKDD), 2009, pp 907–916
9. Gschwind HW, McCluskey EJ (1975) Design of digital computers, vol 59, no 6. Springer, New York, pp 548

Chapter 6
Dimension Reduction and Storage Optimization Techniques for Distributed and Big Data Cluster Environment

S. Kalyan Chakravarthy, N. Sudhakar, E. Srinivasa Reddy,
D. Venkata Subramanian and P. Shankar

Abstract Big Data inherits dimensionality as one of the important characteristics. Dimension reduction is a complex process which aims at converting the dataset from many dimensions to a few dimensions. Dimension reduction and compression techniques are very useful to optimize the storage. In turn, it improves the performance of the cluster. This review paper aims to review different algorithms and techniques which are related to dimensionality reduction and storage encoding. This paper also provides the directions on the applicability of the suitable methodology for Big Data and distributed clusters for effective storage optimization.

Keywords IoT · Sensors · Big Data · Data compression · Dimensionality reduction · Storage · Encoding · PCA · Erasure

---

S. K. Chakravarthy (✉)
Department of Computer Science, University College of Engineering & Technology,
Acharya Nagarjuna University, Guntur, India

N. Sudhakar
Department of Computer Science, Bapatla Engineering College, Bapatla, India

E. S. Reddy
University College of Engineering & Technology, Acharya Nagarjuna University, Guntur,
India

D. V. Subramanian
School of Computer Science, Hindustan Institute of Technology & Science, Chennai, India

P. Shankar
Aarupadai Veedu Institute of Technology, Vinayaka Mission's Research Foundation,
Chennai, India

© The Author(s) 2019   47
N. B. Muppalaneni et al., *Soft Computing and Medical Bioinformatics*,
SpringerBriefs in Forensic and Medical Bioinformatics
https://doi.org/10.1007/978-981-13-0059-2_6

6.1 Introduction

Big Data is a rapidly changing and growing technology in the industries, which creates value for the business in many aspects. Hadoop is an open-source distributed storage and processing framework to handle Big Data. Hadoop distributed le system approach is used to store chunk with threefold technique by default. Storage optimization plays a prominent role when dealing with Big Data and distributed clusters. There are many standard methods such as decision tree, Random Forest, principal component analysis (PCA), and others to reduce dimensions, thereby compressing and reducing the storage space and increasing the computational performance. IoT, cloud, and mobile applications are generating huge data dimensions and subsequent data, which cause not only increase in disk space but also require efficient way of handling the storage. To optimize the storage, dimension reduction, encoding/compression, and or suitable techniques should be adopted. The multilevel storage optimization is one of the architecture options which can help to incorporate both dimension reduction and compression to manage the storage effectively at two levels. Huge data volume in large-scale Big Data environment is primarily due to the huge abnormal data in different format with variety of data types initiated from various sources [1].

This would cause further procurement of storage disks for the data centers with extended networking capability. Increase in size and number of facts are induced not only due to heterogeneity, but also because of the diverse dimensionalities. Therefore, appropriate efforts are required to take place to lessen the volume to effectively examine huge records [2] coming from the data sources. The important consideration for the Big Data is the speed to process large volume of data. There are many IoT applications generating huge volume of data due to large number of sensors and high frequency of data acquisition by the sensors. So, it requires a viable but most effective preprocessing or summarization [3]. The attributes should be effectively reduced to uncover the valuable knowledge [4, 5]. Internet users' behavior can include things like page views, searches, clicks, favorite key words, and selective URLs [6]. One of the key considerations for the storage optimization technique is not only to decrease number of variables or dimensions and at the same time to preserve the important dimensions for better data analytics. The methods to deal with massiveness and heterogeneity of the information should reduce the variable length from multivariable records [7–10]. Cluster deduplication and redundancy elimination algorithms are helpful for elimination of redundant information for supporting efficient output records. These techniques also aimed to provide relevant knowledge patterns through statistics [11, 12]. Social media, sensor networks, scientific experiments, infrastructure, network logs, and the modern gadgets produce a huge amount of useful information. These facts are heterogeneous and always contain multiple formats, multiple sources, aggregated, and include a large record streams.

The nicely designed big information systems need to capable of address all 6Vs correctly via growing stability between information processing goals and the cost of records processing. The processing cost of business record in Big Data environment is the combination of one or more factors such as computational cost, storage cost, programming cost, and others. The primary reasons for the complexity of the records are due to the differences in the formats and dimensions especially inter- and intra-dimensional relationships between various records and attributes. In order to produce useful patterns from available knowledge, there are popular preprocessing techniques like sketching, anomaly detection, dimension reduction, noise removal, outlier detections, and summarization. The primary goal of this preprocessing effort is to reduce the datasets and cleanup obsolete and large facts which are not adding any values to the results. The gathering of irrelevant statistics streams increases the overhead in the computational speed and adds huge cost in terms of IT infrastructure, specially storage systems to handle large volume of data. Therefore, the collection of relevant, and reduced information streams from users is every other mission that calls for critical attention even as designing Big Data systems.

6.2 Dimension Reduction Methods

The following are some of the core benefits of using dimension reduction process in Big Data environment.

1. Efficient data compression, thereby reducing the required storage space.
2. Less computing time due to less number of dimensions.
3. Removal of redundant features.
4. Right validation of the performance of the algorithms and the data models.
5. Reduction of execution of algorithm steps.
6. Improved classification accuracy.
7. Improved clustering accuracy.

The following sections describe some of the popular dimension reduction techniques available which can be helpful for processing complex, heterogeneous data in Big Data environment.

6.2.1 MVR, LVF, HCF, and Random Forests

The best way to reduce dimensions of the data is to carry out using missing values in Missing Value Ratio (MVR) method. Counting the missing values in a dataset can be done either by using statistical models or by using simple database queries and conditions. Columns, which contain many missing values, can be improbable to grasp lots of advantageous information.

One of the simple way of mensuration on how much entropy a data tower has, is to measure its variableness. This is an important step in Low Variance Filter (LVF). In High Correlation Filter (HCF), the pairs of columns with correlation coefficient value larger than a given threshold are reduced to one. Segments with exceptionally practically identical qualities additionally are perhaps to convey fundamentally the same as records.

Random Forest method is also called as decision trees and primarily applied for selecting effective classifiers from the larger functions. The goal is to ensure that every attribute is utilized to find maximum information. Every tree will be trained on a small part of the attributes.

6.2.2 PCA and Multiclass PCA

Principal component analysis (PCA) is a statistic-based tool to map the number of X set of coordinates of a record to a new Y set of coordinates. By applying the PCA procedure against a dataset (m) in the process of dimensionality reduction, more new but small set of m variables are found, while retaining the data transformation information. Support vector machine (SVM) is one of the best feature selection techniques on multivariable data. To improve computational time of feature selection in SVM, it reduces the redundant and non-discriminative features. The multiclass PCA uses both traditional PCA and recursive feature elimination (RFE) methods. RFE is useful to select the required features for performing binary SVM. Each and every time RFE needs to train SVM with large training datasets without PCA dimension reduction. The experimental results from the previous studies show that PCA with PFE is efficient when compared to SVM and other systems. This multiclass PCA helps to produce the efficient dimension reduction for better analytics. The top-ranked features for each pairwise classes and number of principal components with multiclass PCA give accuracy as well as reliable and better results.

6.2.3 BFE and FFC

In backward feature elimination (BFE) method, each iteration can involve the classification algorithm to train on a given N inputs or features. Removal of one feature at a time to train the data model N times on $N - 1$ features. This method begins with all possible n input features until only one last feature is available for classification. Each iteration i provides a model trained on $n - i$ features with an error rate (i). Forward feature construction (FFC) is the inverse system to the backward characteristic elimination. We begin with one function and keep adding one function at a time to produce the best performance. When compared with all other dimension reduction techniques, both PCA and multiclass PCA are found to be more suitable for Big Data clusters. Between PCA and multiclass PCA, multiclass PCA has better data reduction rate percentage of 99 when compared with PCA with 62 and 98% of accuracy when compared with PCA which has 74%. The area under the curve percentage for multiclass PCA is 90% whereas for the standard PCA is 72%.

6.3   Data Compression Methods

Storing the raw data in the Hadoop distributed le system and processing those leads to several challenges. In Hadoop framework, there are numerous difficulties and demanding situations in handling vast informational indexes. Irrespective of whether the facts are stored in HDFS, the fundamental check is that big statistics volumes can motive I/O and gadget-associated bottlenecks. In records escalated Hadoop workloads, I/O operation and device records goes through multiple changes. These changes further delay the completion of the assigned tasks. In enlargement to this inward MapReduce "Rearrange" process is likewise beneath massive I/O weight as it needs to regularly "spill out" center of the street records to close by circles previously progressing from Map degree to reduce phase. For any Hadoop bunch Disk I/O and system transmission capability are taken into consideration as a treasured asset, which need to be disbursed therefore. The record compression in Hadoop framework is often a trade-off among I/O and speed of computation. Compression occurs when the MapReduce reads the facts or while it writes it out. While MapReduce job is red up against compressed records, CPU usage commonly will increase as statistics should be decompressed before the documents can be processed via the Map and Reduce obligation.

Decompression normally increases the time of the job. However, it has been found that during many cases, universal task overall performance improves while compression which is enabled in a couple of stages of activity configuration. Hadoop framework helps many compression formats for both input and output information. A compression format or a codec (compression/decompression) is a set of compiled, ready-to-apply Java libraries that a developer can invoke programmatically to carry out records compression and decompression in MapReduce activity. Each and every codec implements set of rules for compression and decompression and also has one-of-a-kind characteristics.

If the document has been compressed via the use of algorithms that can be split, then information blocks may be decompressed in parallel by way of using numerous MapReduce obligations. However, if the document has been compressed via an algorithm that cannot be split, then Hadoop should pull up blocks collectively and use a single MapReduce mission to decompress them. Some of the compression techniques such as BZIP2, LZO, LZ4, Snappy, and Erasure coding are described in the below sections.

In BZIP2 compression technique, facts are split into Hadoop clusters, and it generates a higher compression ratio than regular GZIP. BZIP2 offers exquisite compression performance, however can be drastically slower than different compression formats together with Snappy in terms of processing performance. The Lempel-Ziv-Oberhumer (LZO) compression layout consists of many smaller (256 K) blocks of compressed facts, allowing jobs to be break up along block barriers. It supports splittable compression, which allows parallel processing of

Table 6.1 XOR (Exclusive OR) operations

X	Y	X ^ Y
0	0	0
0	1	1
1	0	1
1	1	0

compressed textual content le splits via MapReduce jobs. LZO is similar to Snappy in that it is optimized for velocity in place of size. In contrast to Snappy, LZO compressed les can be split, but this calls for a further indexing step. This makes LZO an amazing choice for things like simple-textual content les that are not being stored as part of a box format.

LZ4 is a technique which supports lossless compression. LZ4 library is provided as open-source software using a BSD license. Snappy is one of the latest techniques with APIs to provide both compression and decompression features. This has some restrictions as it does not provide maximum compression and has compatibility issues with other methods and frameworks. It is widely used in many Google Projects.

Erasure coding is a redundant fault-tolerant system. These codec operates on equally sized chunks. The output of the coding technique is parity chunk, and input is an data chunk. The process of generating the parity chunk by taking data chunks is called encoding. The chunks both data and parity chunks are called Erasure coding group. In case of failure, these data chunks can be computed with the help of parity chunk. The Erasure coding is based on exclusive OR. The operations of XOR gate are shown in Table 6.1. When both inputs are 0, then the output is zero. If both inputs are complement, then only the output is high. If the inputs are 1, even then the output is 0. This XOR gate follows associative law, means $X \, Y \, Z = (X \, Y) \, Z$ are equal. XOR gate can take any number of data chunks and produces only one parity chunk. So that it can sustain only one data chunk/disk failure. In case of more than one disk fail, the data cannot be recovered.

In Hadoop distributed file system, the data nodes are built with commodity hardware. The possibility of failures is more common, in the case of hard disks in distributed disk systems. So, the simple XOR technique is not suitable for the HDFS file systems. In order to recover data in case of multiple disk failures, Erasure coding with Reed–Solomon (RS) approach is very useful. Reed–Solomon is configured with multiparity technique with two parameters k, m. This technique accepts a vector with chunks of k and generate a matrix, namely GT to produce the required codeword vector with two chunks. The first one is for data (k) and another one is for parity (m). It can recover data in case of multiple failures by multiplying the GT matrix with the survived storage. This technique aims to tolerate m data chunk failures (Fig. 6.1).

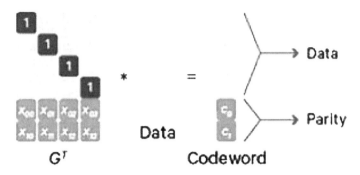

Fig. 6.1 RS with four data chunks and two parity chunks

Table 6.2 Replication of XOR and RS with FT and storage efficiency

	Data durability	Storage efficiency (%)
Single replication	0	100
Three-way replication	2	33
XOR with six data cells	1	86
RS (6,3)	3	67
RS (10,4)	4	71

In the case of Reed–Solomon, the administrator has freedom to choose data durability and storage cost with different k value and m value. This model relies on the number of disk failures due to the m value of parity cell.

$$\frac{k}{k+m}$$

In case of RS configurations with RS (6,3) and RS (10,4), it gives better durability and greater storage efficiency when compared to threefold technique. The RS configuration tolerates even four disk failures with less than 50% overhead in the storage space. The following Table 6.2 shows the comparison of the replication of XOR and RS with FT and storage efficiency.

6.4   Conclusion

Big Data and distributed clusters pose challenges in dealing with storage systems, especially optimal way of managing the storage space as well as the reduction in size. The methods mentioned in this paper are helpful to address the storage issues. The provided literature assessment exhibits that the dimension reduction techniques and compression techniques are very helpful in processing huge information, which is complex in nature. The data dimension reduction should focus primarily on quantity (size) and range (number of dimensions). Using multiclass PCA, the dimension reduction can be efficiently carried out in the data preprocessing phase. To further optimize the storage in Big Data streams, hybrid techniques are required. In general, compression techniques can assist to reduce storage overheads in distributed systems. In this approach, Erasure coding is one of the best techniques to optimize the storage and reduce the storage overhead three times lesser than original. There are different approaches available to compress data, but these techniques are suffering with decompression overhead. To enhance the compressions, intelligent Erasure coding technique is necessary, which can be made as Disk I/O- and network I/O-aware encoding technique. Further research works can be carried out to introduce fairness in disk scheduling and also incorporation of intelligence Erasure coding, reliable and optimal way of handling name node replicas in data nodes for effective storage optimization.

References

1. Pamies-Juarez L, Datta A, Oggier F (2013) Rapidraid: pipelined erasure codes for fast data archival in distributed storage systems. In: INFOCOM, 2013 Proceedings IEEE, IEEE, 2013, pp 1294–1302
2. Thusoo A, Shao Z, Anthony S, Borthakur D, Jain N, Sen Sarma J, Murthy R, Liu H (2010) Data warehousing and analytics infrastructure at face-book. In: Proceedings of the 2010 ACM SIGMOD international conference on management of data, ACM, 2010, pp 1013–1020
3. Li J, Li B (2013) Erasure coding for cloud storage systems: a survey. Tsinghua Sci Technol 18(3):259–272
4. Slavakis K, Giannakis GB, Mateos G (2014) Modeling and optimization for big data analytics: (statistical) learning tools for our era of data deluge. IEEE Signal Process Mag 31 (5):18–31
5. Miller JA, Bowman C, Harish VG, Quinn S (2016) Open source big data analytics frameworks written in scala. In: IEEE international congress on big data (BigData Congress), IEEE, 2016, pp 389–393
6. Zhang Z, Deshpande A, Ma X, Thereska E, Narayanan D (2010) Does erasure coding have a role to play in my data center. Microsoft research MSR-TR-2010 52
7. Wang A, G, U. M, Wang A, Zheng K, G, U. M. (2015). Cloudera Engineering Blog Introduction to HDFS Erasure Coding in Apache Hadoop, (6).
8. Shahabinejad M, Khabbazian M, Ardakani M (2014) An efficient binary locally repairable code for hadoop distributed le system. IEEE Commun Lett 18(8):1287–1290
9. Khan O, Burns RC, Plank JS, Pierce W, Huang C (2012) Rethinking erasure codes for cloud le systems: minimizing i/o for recovery and degraded reads. In: FAST, 2012, p 20
10. Blaum M, Brady J, Bruck J, and Menon J Evenodd: An efficient scheme for tolerating double disk failures in raid architectures. IEEE Transactions on Computers, 44(2):192–202, 1995
11. Rashmi K, Shah NB, Gu D, Kuang H, Borthakur D, Ramchandran K (2015) A hitchhiker's guide to fast and efficient data reconstruction in erasure-coded data centers. ACM SIGCOMM Comput Commun Rev 44(4):331–342
12. Fan B, Tantisiriroj W, Xiao L, Gibson G (2009) Diskreduce: raid for data intensive scalable computing. In: Proceedings of the 4th annual workshop on petascale data storage, ACM, 2009, pp 6–10

Chapter 7
A Comparative Analysis on Ensemble Classifiers for Concept Drifting Data Streams

Nalini Nagendran, H. Parveen Sultana and Amitrajit Sarkar

Abstract Mining in data stream plays a vital role in Big Data analytics. Traffic management, sensor networks and monitoring, weblogs analysis are the application of dynamic environments which generate streaming data. In a dynamic environment, data arrives at high speed and algorithms that process them need to fulfill the constraints on limited memory, computation time, and one-time scan of incoming data. The significant challenge in data stream mining is data distribution changes over a time period which is called concept drifts. So, learning model need to detect the changes and adapt according to that model. By nature, ensemble classifiers are adapting to changes very well and deal the concept drift very well. Three ensemble-based approaches were used to handle the concept drift: online, block-based ensemble, and hybrid approaches. We provide a survey on various ensemble classifiers for learning in data stream mining. Finally, we compare their performance on accuracy, memory, and time on synthetic and real datasets with different drift scenarios.

Keywords Data mining · Data stream · Sudden drift · Gradual drift and ensemble classifier

7.1 Introduction

In recent days, applications such as traffic information system [1], sensor networks [2], electricity filtering [3], intrusion detection [4], and credit card fraud detection [5], which generate data continuously, are known as data stream [6]. In a dynamic environment, data arrives at high speed and algorithms that process them need to fulfill the constraints on limited memory, computation time, and one-time scan of

N. Nagendran (✉) · H. P. Sultana
School of Computer Science and Engineering, Vellore Institute
of Technology, Vellore, Tamil Nadu, India

A. Sarkar
Department of Computing, Ara Institute of Canterbury,
Christchurch, Central City, New Zealand

© The Author(s) 2019   55
N. B. Muppalaneni et al., *Soft Computing and Medical Bioinformatics*,
SpringerBriefs in Forensic and Medical Bioinformatics
https://doi.org/10.1007/978-981-13-0059-2_7

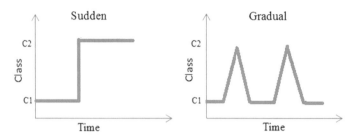

Fig. 7.1 Sudden and gradual drifts

incoming data. Traditional data mining approaches are not capable of dealing continuously arriving data and also less accurate while classifying the data in the data stream. Mining in the data stream is challenging task because data arrives continuously, and algorithm scans the incoming data at one time with the limited amount of memory.

The major challenge in data stream mining is data distribution changes over a time period which is called concept drifts [7]. For example, user consumption of food product in the supermarket and customer interest toward fashion may change over time due to the season, festivals, economy, and so on. These changes affect the accuracy of the learning model drastically. Therefore, classification algorithm needs to be updated according to the changes in the incoming data. According to the speed of changes, concept drifts are classified into two types [8], sudden (abrupt) and gradual drifts. In sudden drift, changes happen from one concept to another concept abruptly, e.g., seasonal changes on sale. In gradual drift, changes happen incrementally between concepts, e.g., dehydration of sensors due to temperature and humidity. Figure 7.1 shows sudden drift changes enormously between the underlying class distribution and the incoming data in a short period of time, while gradual drift is changed happen slowly between the underlying class distribution and the incoming data.

Many surveys [9, 10] have been proposed to handle the concept drift. But that survey was missed to compare hybrid approaches. In this paper, we have discussed ensemble classifier. Ensemble methods are mainly used in concept drift of evolving data stream. There are three types approaches used in ensemble classifier, such as online ensembles [11, 12], block-based ensembles [13–16], and hybrid approaches [17, 18].

7.2 Ensemble Classifier on Data Stream

Ensemble classifier is composite models that are used a combination of classifiers whose decisions are aggregated by a voting mechanism.

This ensemble handles concept drift very efficiently. They are more accurate than single classifier [19] because composite model reduces the variance of the individual classifier. The ensemble classifier accuracy can be varied by data, attributes, and base learner/classifier because class label prediction is established by member voting on incoming instances. Online ensemble, block-based ensembles, and hybrid approaches are discussed in this paper.

7.2.1 Online Ensembles

Online ensembles update component weights after each instance without the need for storage and reprocessing. This approach can adapt to sudden changes as earlier as possible.

Weighted majority algorithm [20] was the first proposed algorithm on the online ensemble. It combines the predictions of a set of the component classifier and updates their weights when they make a false prediction.

Oza have developed online bagging and online boosting [11]. These algorithms combine the weaker learner to create strong learner that make an accurate prediction. Online bagging creates bootstrap samples of a training set using sampling with replacement. Each bootstrap sample is used to train a different component of the base classifier. The online boosting method used weighting procedure of AdaBoost which divides the instances into two portions; half of the weight is assigned to correctly classified instances, and other half is assigned to misclassified instances. Online bagging is more accurate than online boosting which is than adding weight into misclassified instances rather than adding more random weight to all instances. Though bagging is more robust to noise than boosting, it is better than bagging. Bagging does not perform pruning and reweights on component periodically. Hence, the bagging may increase the computational cost and not react to gradual changes.

Leverage bagging [12] is a new version of online bagging which adds more randomization in input and increases the output of the classifier. This algorithm reacts faster to sudden drift. The main drawbacks of the algorithm are not performing pruning periodically which increases the computational cost and not respond to gradual drift. Online ensembles are the best approaches to improving the accuracy.

7.2.2 Block-Based Ensembles

In block-based approaches, instances arrive in portions, called as blocks. Most block-based ensembles evaluate their components periodically and replace the weakest ensemble member with a new classifier. This approach reacts very well to gradual concept drift rather than sudden drift and ensures accuracy in it.

Stream ensemble algorithm (SEA) [13] was the first algorithm on block-based ensemble approaches. It processes the stream divided into continuous fixed blocks that are called chunks. After processing each block by the learning algorithm, SEA evaluates score on the new block, and the weakest classifier can be replaced by a newly trained one. The overall prediction of SEA is measured by a majority vote of the prediction of the classifier in the ensemble. SEA is a batch-based algorithm which makes concept fluctuation. So it is lack to respond abrupt changes which may be lead to some unpredictable factors in a data stream.

Accuracy weighted ensemble (AWE) [14], the authors proposed to train a new classifier on each incoming data chunk by base classifier. Then, existing component classifiers in the ensemble are evaluated on the most recent chunk. A special version of mean square error (MSE) was used to evaluate algorithms which chose the k best classifier to create a new ensemble. The prediction of AWE was based on a weighted majority vote of the classifier in the ensemble. The formula is given below which proposed by Wang et al.,

- Mean squared error of classifier C_i can be expressed by

$$\text{MSE}_i = \frac{1}{|S_n|} \sum_{(x,n \in S_n)} \left(1 - f_c^i(x)\right)^2 \tag{7.1}$$

- A classifier which predicts randomly will have mean square error which randomly predicts the error classifier

$$\text{MSE}_r = \sum_c p(c)(1 - p(c))^2 \tag{7.2}$$

- Weight w_i for classifier C_i from (7.1) and (7.2)

$$w_i = \text{MSE}_r - \text{MSE}_i \tag{7.3}$$

where S_n consists of records in the form of (x, c), where c is the true label of the record, C_i classification error of example (x, c) is $1 - f_c^i(x)$, where $f_c^i(x)$ is the probability given by C_i that x is an instance of class c and $p(c)$ probability of x being classified as class c. AWE increased the accuracy when compared with a single classifier. The drawback of the algorithm is that (i) it is not possible to remember all the components due to limited memory and (ii) the accuracy depends upon on the block size.

Brzeinski et al. proposed an algorithm Accuracy Updated Ensemble (AUE) [15] and AUE Version 2 (AUE2) [16] to provide more accuracy. The AUE2 was motivated by AWE; it improves efficiency and reduces the computational cost. The AUE2 processes the stream of data as a block-based ensemble with incremental nature of Hoeffding tree. The main disadvantage of AUE2 is block size which may reduce the accuracy of the algorithm. Because, larger blocks may cause the delay of the adaptation to a new concept, while a too small block is not sufficient enough to build a new classifier.

7.2.3 Hybrid Approaches

Hybrid approaches were combining the best characteristic of online and block-based ensemble classifiers to produce new ensemble classifier. These approaches produced more accuracy than other two ensembles. Brzeinski et al. proposed a new incremental classifier, referred to as online accuracy updated ensemble (OAUE) [17], which trains and weights element classifiers with every incoming example. The primary novelty of the OAUE algorithm is weighting mechanism, which estimates a classifier's blunders on a window of remaining seen times in regular time and reminiscence without the need of remembering previous examples. They analyzed which strategies for reworking block-based ensembles into online beginners are most promising in phrases of type accuracy and computational expenses.

Adaptive window online ensemble (AWOE) [18] was an online ensemble with internal change detector which proposed by Sun et al., which keeps a pool of weighted classifiers by using obtaining the very last output of components primarily based on the weighted majority voting rule. The sliding window is selected to reveal the classification error of the most current records. Moreover, an extended-term buffer mechanism is chosen to keep the current training instances, on which a new classifier is constructed while a change is detected.

Moreover, the addition of an online learner and go with the flow detector gives faster reactions to surprising concept modifications compared to most block-based ensembles. Such method ensures that the maximum current data is protected in the final prediction. They undertake an incremental set of rules for building decision trees, which is known as Hoeffding tree. It builds a decision tree from statistics streams incrementally, without storing instances after they were employed to renew the tree.

7.3   Results and Discussions

7.3.1   Comparison of Ensemble Classifiers

The performance of data stream classification measured in three dimensions includes time, memory, and accuracy [21, 22]. The online ensemble takes more time to process the data, but it is providing more accuracy than block-based approaches. In contrast to online bagging, members of block-based ensembles are weighted, can be removed, and not always updated on the algorithm. The main disadvantage of block-based ensembles is the difficulty of defining block size. Hybrid approaches solve all the above-mentioned problems.

7.3.2   Experimental Results

Massive online analysis (MOA) [23, 24] tool is used to analyze the various ensemble classifiers. In this study, we compare two online ensemble classifiers, two block-based classifiers, and hybrid approaches. We used two synthetics such as

Table 7.1 Characteristics of dataset

Dataset	No. of instances	Attributes	Classes
SEA	1 M	3	4
LED	1 M	24	10
CoverType	581 K	53	7
Electricity	45 K	7	2

Table 7.2 Accuracy of ensemble classifiers (%)

	SEA	LED	CoverType	Electricity
Ozabag	88.8	**67.62**	80.4	77.34
Levbag	87.35	54.99	89.13	91.45
AWE	79.59	52.29	78.74	70.84
AUE2	80.81	53.41	88.14	77.34
OAUE	**88.83**	53.43	90.56	87.93
AWOE	89.12	57.78	**95.26**	**88.96**

Table 7.3 Processing Time of Ensemble Classifiers (Seconds)

	SEA	LED	CoverType	Electricity
Ozabag	70.54	**20.1**	250.6	11.38
Levbag	81.49	34.34	876.61	38.48
AWE	**10.75**	53.54	338.94	14.94
AUE2	11.64	43.35	130.42	10.03
OAUE	14.24	36.55	**107.11**	**8.3**
AWOE	12.62	33.23	221.8	18.94

Table 7.4 Memory Usage of Ensemble Classifiers (MB)

	SEA	LED	CoverType	Electricity
Ozabag	1.12	1.5	6.45	0.64
Levbag	69.45	1.95	7.15	0.66
AWE	**0.71**	0.61	3.12	0.91
AUE2	1.76	**0.22**	1.05	0.24
OAUE	5.99	0.72	3.12	1.55
AWOE	1.14	0.23	**0.6**	**0.63**

SEA and LED which are available in the MOA [23, 24], respectively, and two real-world datasets such as CoverType and Electricity which are available in the UCI repository [25], Table 7.1 shows dataset characteristics which are number of instances, attributes, and classes.

As Tables 7.2, 7.3, and 7.4 depict, the average prequential accuracy, average processing time, and memory usage of the different ensemble classifiers.

In real-world dynamic environment [26, 27], conceptual changes have unpredictability and uncertainty which could better verify the performance of the algorithm. Figure 7.2 depicts the accuracy changes on the CoverType. AWOE produced

Fig. 7.2 Accuracy on
ensemble classifiers

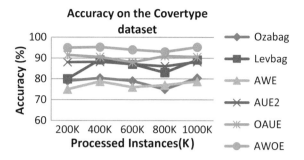

the best result, followed by the OAUE. AWOE accuracy curve is stable and robust to concept drift, which additionally suggests that AWOE algorithm has higher adaptability for actual environment.

7.4 Conclusions

This paper discusses various ensemble approaches to handling concept drifting data stream. We experiment accuracy, memory usage, and processing time of these approaches on MOA interfaces and then compare the performance of these algorithms. Hybrid method has better performance than other two ensembles based on the following three elements: (i) It resolves the hassle of setting an appropriate length of block; (ii) it handles different type of drift very well; (iii) the hybrid approach is more efficient than different ensemble strategies in terms of accuracy and memory usage.

From this study, we conclude that data stream mining is in early stages and still more challenges to be solved. So in future, we plan to design a new ensemble classifier for real big data application which increases the accuracy, reduce the processing time and memory usage.

References

1. Geisler S, Quix C, Schiffer S, Jarke M (2012) An evaluation framework for traffic information systems based on data streams. Transp Res Part C: Emerg Tech 23:29–55
2. Cohen L, Avrahami-Bakish G, Last M, Kandel A, Kipersztok O (2008) Real-time data mining of non-stationary data streams from sensor networks. Inf Fusion 9(3):344–353
3. Delany SJ, Cunningham P, Tsymbal A, Coyle L (2005) A case-based technique for tracking concept drift in electricity filtering. Knowl-Based Syst 187–188
4. Lane T, Brodley CE (1998) Approaches to online learning and concept drift for user identification in computer security. In: The fourth international conference on knowledge discovery and data mining—KDD-98, 1998, pp 259–263
5. Garofalakis M, Gehrke J, Rastogi R (2002) Querying and mining data streams: you only get one look a tutorial. In: Proceedings of the 2002 ACM SIGMOD international conference on management of data, Madison, WI, USA, 2002, p 635

6. Aggarwal CC (2007) Data streams: models and algorithms, vol 31. Springer Science and Business Media, Kluwer Academic Publishers, London, pp 1–372
7. Widmer G, Kubat M (1996) Learning in the presence of concept drift and hidden contexts. Mach Learn 23(1):69–101
8. Tsymbal A (2004) The problem of concept drift: definitions and related work, vol 106. Computer Science Department, Trinity College, Dublin, Ireland, Tech. Rep. 2004, pp 1–7
9. Ditzler G, Roveri M, Alippi C, Polikar R (2015) Learning in nonstationary environments: a survey. IEEE Comput Intell Mag 10(4):1–14
10. Gama J, Zliobaite I, Bifet A, Pechenizkiy M, Bouchachia A (2014) A survey on concept drifts adaptation. ACM Comput Surv 46(4):1–44
11. Oza NC (2005) Online bagging and boosting. In: 2005 IEEE International conference on systems, man and cybernetics, vol 3, Waikoloa, HI, USA, pp 2340–2345
12. Bifet A, Holmes G, Pfahringer B (2010) Leveraging bagging for evolving data streams. In: Joint European conference on machine learning and knowledge discovery in databases, Barcelona, Spain, pp 135–150
13. Street WN, Kim Y (2001) A streaming ensemble algorithm (SEA) for large-scale classification. In: Proceedings of the 7th ACM SIGKDD international conference on Knowledge discovery and data mining, San Francisco, CA, USA, pp 377–382
14. Wang H, Fan W, Yu PS, Han J (2003) Mining concept-drifting data streams using ensemble classifiers. In: Proceedings of the ninth ACM SIGKDD international conference on Knowledge discovery and data mining, Washington, DC, USA, pp 226–235
15. Brzezinski D, Stefanowski J (2011) Accuracy updated ensemble for data streams with concept drift. In: 6th International conference on hybrid artificial intelligence systems, Wroclaw, Poland, pp 155–159
16. Brzezinski D, Stefanowski J (2014) Reacting to different types of concept drift: the accuracy updated ensemble algorithm. IEEE Trans Neural Networks Learn Syst 25(1):81–94
17. Brzezinski D, Stefanowski J (2014) Combining block-based and online methods in learning ensembles from concept drifting data streams. Inf Sci 265:50–67
18. Sun Y, Wang Z, Liu H, Du C, Yuan J (2016) Online ensemble using adaptive windowing for data streams with concept drift. Int J Distrib Sens Netw 12(5):1–9
19. Maimon O, Rokach L (2010) Data mining and knowledge discovery handbook. Springer Science & Business Media, London, pp 1–1306
20. Littlestone N, Warmuth MK (1994) The weighted majority algorithm. Inf Comput 108 (2):212–261
21. Gama J, Sebastião R, Rodrigues P (2013) On evaluating stream learning algorithms. Mach Learn 90(3):317–346
22. Bifet A, Francisci Morales G, Read J, Holmes G, Pfahringer B (2015) Efficient online evaluation of big data stream classifiers. In: Proceedings of the 21th ACM SIGKDD international conference on knowledge discovery and data mining, Sydney, NSW, Australia, pp 59–68
23. Bifet A, Holmes G, Kirkby R, Pfahringer B (2010) Moa: massive online analysis. J Mach Learn Res 11:1601–1603
24. Bifet A, Holmes G, Pfahringer B, Kranen P, Kremer H, Jansen T, Seidl T (2010) MOA: massive online analysis, a framework for stream classification and clustering. In: Proceedings of the first workshop on applications of pattern analysis, Cumberland Lodge, Windsor, UK, pp 44–48
25. Frank A, Asuncion A (2010) UCI machine learning repository. http://archive.ics.uci.edu/ml
26. Krempl G, Žliobaite I, Brzeziński D, Hüllermeier E, Last M, Lemaire V, Stefanowski J (2014) Open challenges for data stream mining research. ACM SIGKDD Explor Newsl 16(1):1–10
27. Krawczyk B, Stefanowski J, Wozniak M (2015) Data stream classification and big data analytics. Neurocomputing 150(PA):238–239

Chapter 8
Machine Learning Algorithms with ROC Curve for Predicting and Diagnosing the Heart Disease

R. Kannan and V. Vasanthi

Abstract Heart diseases are now becoming the leading cause of mortality in India with a significant risk of both males and females. According to the Indian Heart Association (IHA), four people die of heart diseases every minute in India and the age-groups are mainly between 30 and 50. The one-fourth of heart failure morality occurs to people less than 40. A day in India nine hundred people dies below the age of 30 due to different heart diseases. Therefore, it is imperative to predict the heart diseases at a premature phase with accuracy and speed to secure the millions of people lives. This paper aims to examine and compare the accuracy of four different machine learning algorithms with receiver operating characteristic (ROC) curve for predicting and diagnosing heart disease by the 14 attributes from UCI Cardiac Datasets.

Keywords Machine learning algorithms · Gradient boosting · ROC curve
Heart disease

8.1 Introduction

The explosive growth of health-related data presented unprecedented opportunities for improving health of a patient. Heart disease is the dominant reason for mortality in India, Australia, UK, USA, and so on. Machine learning involves and activates the uncovering new trends in healthcare industries. By using machine learning technique we can conduct the research from different aspects between heart diseased persons and healthy person based on their existing medical considerable datasets. Tremendous approach in this study of all cardiac-related disease classification is done to find the

R. Kannan · V. Vasanthi (✉)
Department of Computer Science, Rathinam College of Arts & Science,
Coimbatore, Tamil Nadu, India
e-mail: vasanthi.cs@rathinam.in

R. Kannan
e-mail: dschennai@outlook.com

© The Author(s) 2019
N. B. Muppalaneni et al., *Soft Computing and Medical Bioinformatics*,
SpringerBriefs in Forensic and Medical Bioinformatics
https://doi.org/10.1007/978-981-13-0059-2_8

63

disguised medical information. It accelerated the establishment of vital knowledge, e.g., patterns, different dimensions for identifying relationships amidst medical factors interconnected with heart diseases.

By using some machine learning techniques, heart disease prediction can be made simple by using various characteristics to find out whether the person suffers from heart attack or not, and it also takes less time to predict and improve the medical diagnosis of diseases with good accuracy and minimizes the occurrence of heart attack. It assists to resolve the hidden reason and diagnose the heart diseases efficiently even with the uncertainties and inaccuracies. This paper emphasizes the machine learning algorithms such as logistic regression, Random Forest, boosted tree, stochastic gradient boosting and support vector machines that are used to confirm the best prediction technique in terms of its accuracy and error rate on the specific dataset.

8.1.1 Heart Diseases

Heart diseases are life-frightening diseases, and it should be contemplating as a global health precedence. Moreover, heart diseases reside a great stress on patients, caretaker, and healthcare systems. For the present, almost 30 million people worldwide are living with the heart diseases such as patients affected by heart dieses because of cholesterol deposits, high blood sugar, poor in hygiene, physically inactive, unhealthy diet, smoking and hand change smoking, being overweight, high blood pressure, viral infection compared with the survival rates worse than any other diseases. Although the survival which is globally 48.9 million, 50 people affected by heart defects by birth defects. The signs and symptoms of 51 heart diseases mentioned below are caused because of above highlighted major reasons.

- Chest pain
- Shortness of breath
- Sweating
- Nausea
- Irregular heartbeat
- Throat or jaw pain
- A cough that won't quit.

8.1.2 Machine Learning (ML)

Machine learning is the technique to allow the computers to learn and predict automatically for achieving the difficult jobs whose processes cannot be simply described by humans, for example, self-driving car, Netflix movies rating, Amazon online recommendations, diagnosis in medical imaging, autonomous robotic

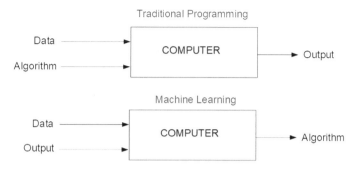

Fig. 8.1 Overview of machine learning

surgery. The machine learnings are of three different types such as supervised learning, unsupervised learning, and reinforcement learning. The following Fig. 8.1 shows the contradiction between the traditional programming and machine learning to provide the best results.

8.1.3 Supervised Machine Learning Algorithms

Supervised learning is the one type of machine learning algorithms used to learn and predict with labeled training data. The training data contains the set of training examples, and each example has the input object and the desired output value. The supervised learning algorithms examine the training data, produce the complete function, which can be used for mapping new examples, and correctly determine the class labels for hidden examples. Here the below list shows the supervised machine learning algorithms [1].

- Linear regression
- Logistic regression
- Decision tree
- Support vector machine (SVM)
- Random Forest
- And so on.

8.1.4 Unsupervised Machine Learning Algorithms

In contrast, unsupervised machine learning algorithms are used to learn and predict without labeled training data. Here the below list shows the unsupervised machine learning algorithms [1].

- Clustering
- Anomaly detection
- Approaches for learning latent variable models
- And so on.

8.1.5   Reinforcement Machine Learning Algorithms

Reinforcement machine learning algorithms differ from supervised machine learning and unsupervised machine learning. Reinforcement machine learning algorithms allow technologies and software agents to automatically determine the ideal behavior within a specific context in order to maximize its performance. Simple reward response is required for the agent to learn which action is best; this is known as the reinforcement signal [2].

8.2   Materials and Methods

In this section, we have introduced the heart disease Cleveland dataset and explained the four different machine learning models with ROC curve for the heart disease predictions with diagnoses.

8.2.1   The Cleveland Dataset

Machine learning models need a certain amount of data to lead an adequate algorithm. We can collect the necessary datasets from repository of healthcare industries and third-party data sources. It is enabling comparative effectiveness in the research done by producing unique and powerful machine learning algorithms [3].

In this paper, we have obtained 303 records with 14 set of variables and divided the data into training (70%) and testing (30%) from the Cleveland dataset. Percentage of heart disease should not be the same in training and testing data. We have listed the 14 attributes (Table 8.1) below.

The variable we want to predict is Num with value 0: <50% diameter narrowing and value 1: >50% diameter narrowing. We assume that every value with 0 means heart is normal for patient and 1,2,3,4 means heart disease.

Table 8.1 Heart disease attributes

Variable name	Description
Age	Age in years
Sex	Sex, 1 for male, 0 for female
CP	Chest pain type (1 = typical angina; 2 = atypical angina; 3 = non-angina pain; 4 = asymptomatic)
Trestbps	Resting blood pressure
Chol	Serum cholesterol in mg/dl
Fps	Fasting blood sugar larger 120 mg/dl (1 true)
Restecg	Resting electrocardiographic results (1 = abnormality, 0 = normal)
Thalach	Maximum heart rate achieved
Exang	Exercise-induced angina (1 yes)
Oldpeak	ST depression induce. Exercise relative to rest.
Slope	Slope of peak exercise ST
CA	Number of major vessel
Thal	No explanation provided, but probably thalassemia
Num	Diagnosis of heart disease (angiographic disease status) 0 (<50% diameter narrowing) 1 (>50% diameter narrowing)

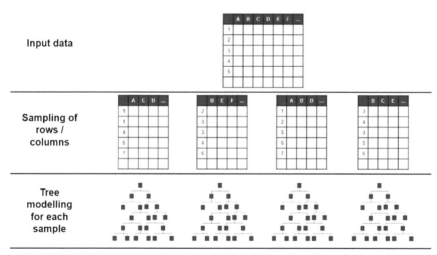

Fig. 8.2 Decision tree and Random Forest

8.2.2   Random Forests

Random Forest is a supervised machine learning algorithm. We can see it from its name, which creates a forest by random tree. It contains a direct relationship between the number of trees in the forest (Fig. 8.2 shows the relationship) and the

Fig. 8.3 Logistic regression

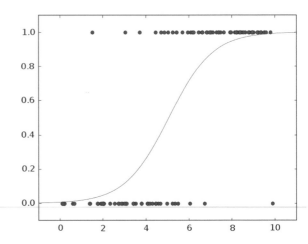

results it can get [4]. It can be used for both classification and regression tasks. Over fitting is one critical problem that may make the results worse, but for Random Forest algorithm, if there are enough trees in the forest, the classifier will not over fit the model. The third advantage is the classifier of Random Forest can handle missing values, and the last advantage is that the Random Forest classifier can be modeled for categorical values.

8.2.3 Logistic Regression

Logistic regression algorithm is a regression and classification method for evaluating the dataset in which it contains one or more independent variables that conclude an outcome. The outcome is measured with a divided variable (in which can be two possible outcomes) [5]. The following Fig. 8.3 shows the two possible outcomes from logistic regression.

8.2.4 Gradient Boosting

Gradient boosting is one of the best supervised machine learning algorithms for regression and classification problems. Gradient boosting algorithm is the form of an ensemble of weak prediction models likely decision trees. It builds the model in a stage-wise fashion like other boosting methods do, and it generalizes them by allowing optimization of an arbitrary differentiable loss function [6]. The following Fig. 8.4 shows the gradient boosting algorithms.

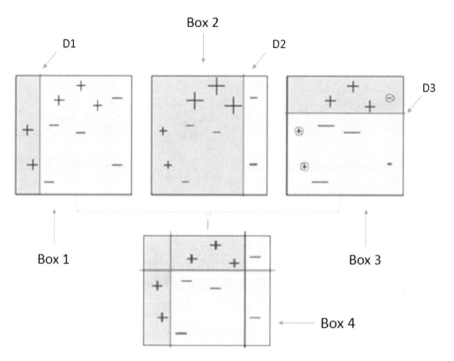

Fig. 8.4 Gradient Boosting

8.2.5 *Support Vector Machine*

Support vector machine (SVM) is a supervised machine learning algorithm. SVM can be used for both the classification and regression problems. However, it is common for classification problems [7]. In this algorithm, we plot each data item as a point in *n*-dimensional space (where *n* is number of features you have) with the value of each feature being the value of a particular coordinate. Then, we perform classification by finding the hyperplane that differentiates the two classes very well (Fig. 8.5 shows an example).

8.2.6 *ROC Curve*

The receiver operating characteristic (ROC) plot is a most popular measure for evaluating classifier performance [7]. The ROC plot is based on two basic evaluation measures—specificity and sensitivity. Specificity is a performance measure of the negative part, and sensitivity is a performance measure of the positive part. The majority of machine learning models produce some kind of scores in addition to

Fig. 8.5 Support vector
machine

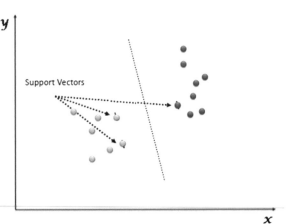

predicted labels. These scores can be discriminant values, posterior probabilities, and so on. Model-wide evaluation measures are calculated by moving threshold values across the scores.

8.2.7 Software Tools

As time goes on, there are best tools and software that are available to predict the heart diseases in the market. Here, we used the most favorable open-source software name called as "R programming" for this paper. R is the most popular statistical computing and visualization software package in the worldwide, which used in number of growing healthcare industries, commercial, government organizations, academics, and research.

8.3 Experimental Evaluation

We have applied the four machine learning algorithms such as logistic regression, Random Forest, stochastic gradient boosting, and support vector machine with the attributes taken from the UCI heart diseases dataset. Eventually, we predicted the four different results from different tuning parameters and compared the best model of each machine learning algorithm with ROC curve. Here Table 8.2 shows the results comparison of ACU and accuracy between models.

The results for the compression of ACU and accuracy between the machine learning models represented as graphical visualization are shown in Table 8.2 and Fig. 8.6.

Table 8.2 Comparison of ACU and accuracy between models

Algorithms	ACU	Accuracy
Logistic Regression	0.9161585	0.8651685
Random Forest	0.8953252	0.8089888
Stochastic gradient boosting	0.9070122	0.8426966
Support vector machine	0.882622	0.7977528

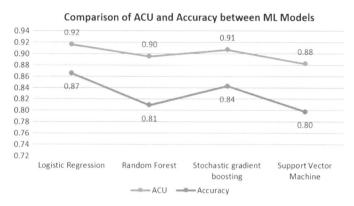

Fig. 8.6 Visualizing the results of ACU and accuracy models

8.4   Results and Conclusion

Fourteen predictor variables from the UCI heart disease dataset are used to predict the diagnosis of heart disease. The performance of the four different machine learning algorithms, such as logistic regression, stochastic gradient boosting, Random Forest, and support vector machines are compared with the accuracy obtained from them.

Thirty percent of the data is hold out as a testing dataset that is not seen during the training stage of the data. During the training of boosted and support vector machines, tenfold cross-validation is used to maximize the ROC (parameter tuning) and select the appropriate models. A comparison of the area under the ROC and the accuracy of the model predictions shows that logistic regression performs best and it can predict with 0.87% of accuracy.

8.5 Future Work

This paper has summarized the methods for heart disease prediction along with innovative machine learning algorithms. As a future work, we have planned to predict the heart diseases using tensor flow of deep learning algorithms with more dataset. This deep learning tensor flow will automate and increase the process of prediction in terms of speed.

References

1. What is Machine Learning? A definition. www.expertsystem.com/machine-learning-definition/
2. Elshaw M, Mayer NM (2008) Reinforcement learning edited by Cornelius Weber
3. Heart Disease Data Set. http://archive.ics.uci.edu/ml/datasets/heart+Disease
4. How Random Forest Algorithm Works in Machine Learning. https://medium.com/@Synced/how-random-forest-algorithm-works-in-machine-learning-3c0fe15b6674
5. The Pennsylvania State University, 'STAT 504 | Analysis of Discrete Data'. https://onlinecourses.science.psu.edu/stat504/node/149
6. Boosting Machines. https://github.com/ledell/useR-machine-learning-tutorial/blob/master/gradient-boosting-machines.Rmd
7. Introduction to the ROC (Receiver Operating Characteristics) plot. https://classeval.wordpress.com/introduction/introduction-to-the-roc-receiver-operating-characteristics-plot
8. Practical Machine Learning, https://www.coursera.org/learn/practical-machine-learning
9. Dr. Brownlee J (2017) Master machine learning algorithms, eBook 2017. https://machinelearningmastery.com
10. Daumé III H, A course in machine learning. https://ciml.info
11. Understanding Support Vector Machine algorithm. https://www.analyticsvidhya.com/blog/2017/09/understaing-support-vector-machine-example-code

Chapter 9
Latent Variate Factorial Principal Component Analysis of Microelectrode Recording of Subthalamic Nuclei Neural Signals with Deep Brain Stimulator in Parkinson Disease

Venkateshwarla Rama Raju, Lavanya Neerati and B Sreenivas

Abstract Parkinson's disease (PD) is a complex chronic disorder characterized by four classes of cardinal motor symptom features—tremor, bradykinesia, rigidity, and postural instability. Even though clinical benefits of deep-brain-stimulator (DBS) in subthalamic-nuclei (STN) neurons have been established, albeit, how its mechanisms enhances motor-features reduces tremor and restores and increases motor function have not been fully customized. Also, its objective methods for quantifying efficacy of DBS are lacking. We present a latent variate factorial (or factor) principal component analysis-based method to predict UPDR score objectively. Twelve PD subjects are included in this study. Our hypothesis in this study is that whether the DBS saves the STN neurons and restores motor function after the reduction of tremor or damages. In our long study, the high-frequency stimulation in diseased brain did not damage subthalamic nuclei (STN) neurons but protected. Further, it is risk-free to stimulate STN much prior than it was accepted far so. The latent variate factorial is a statistical mathematical technique principal component analysis (PCA)-based method for computing the effects of DBS in PD. We extracted and then extrapolated microelectrode signal recordings (MER) of STN

V. R. Raju (✉) · L. Neerati · B. Sreenivas
CMR College of Engineering & Technology (UGC Autonomous), Kandlakoya,
Medchal Road, Hyderabad 501401, Telangana, India
e-mail: drvrr@cmrcet.org

L. Neerati
e-mail: nlavanya12@gmail.com

B. Sreenivas
e-mail: srinu.vagg@gmail.com

V. R. Raju
Nizam's Institute of Medical Sciences, Hyderabad, India

V. R. Raju
Osmania University College of Engineering (Autonomous), Hyderabad, India

© The Author(s) 2019   73
N. B. Muppalaneni et al., *Soft Computing and Medical Bioinformatics*,
SpringerBriefs in Forensic and Medical Bioinformatics
https://doi.org/10.1007/978-981-13-0059-2_9

neurons features. The signal parameters were transformed into a lower-dimensional feature space. So we could predict the disease at an early stage.

Keywords Parkinson's disease (PD) · Deep brain stimulation (DBS) Subthalamic-Nuclei (STN) · Microelectrode recording (MER) · Principal component analysis (PCA)

9.1 Introduction

Parkinson's-disease (PD) is a complex chronic disorder characterized by four classes of cardinal features—tremor, bradykinesia, rigidity, and postural instability [1–4]. Perhaps, PD is best identified for its tremor, slowness and also stiffens-movements.

DBS is a technique that decreases and/or diminishes motor features, restores, and increases motor function in PD subjects. It was formulated by De Long [5] a new model for the brain's circuitry and devised by Alim [6] an efficient reversible—intervention that remedies neural—misfiring. For the development of DBS, the duo bagged a prestigious Medical Discovery Award [5–7] in 2014.

Even though clinical benefits of DBS in STN established, albeit, how its mechanisms improve motor features in PD is not elucidated fully. Hence, a need arises for scientific objective evidence.

The DBS stimulation parameters (stimulation—intensity, frequency, and pulse width, amplitudes) are set by subjective evaluation of symptoms based on the UPDRS scale but not objectively [6]. In this study, we present a latent variate factorial (or factor) mathematical statistical technique principal component analysis (PCA)-based tracking method for computing the effects of DBS in PD to improve the effects of DBS in PD. Using microelectrode recordings (MERs) of STN neural signals acquisition with deep brain stimulator, the PD characteristic signal features were (parameters of nonlinear dynamics) extracted through various parameters (amplitude, pulse width, frequency, RMS, coherence, correlation, and entropy). Using the latent variate factorial by the PCA, original parameters were transformed into a smaller number of parameters considering first principal component (PC) as a first highest magnitude of the eigenvalue corresponding to diagonal eigenvector, i.e., PC1, second PC as the next highest magnitude eigenvector, i.e., PC2, and so on in the order of decreasing. Finally, effects of DBS were quantified by examining the PCs into a lower-dimensional feature space. Twelve subjects were employed in this study after checking their cognitive function. The main hypothesis of our study is that when we give high-frequency stimulations to subthalamic nucleus (STN), it will protect neurons in the substantia nigra and it will improve the features of PDs or destroys. We hypothesize that microelectrode recording (MER) with high-frequency deep brain stimulation (DBS) of the STN protected the neurons of substantia nigra which is an effective method for subjects with Parkinson's disease (PD). However frequent battery changes are one problem and if at all any infection of implantation of electrode [7, 8].

9.2 Objective

The prime objective of this study is to quantify—compute the efficacy of MER with STN DBS by using a latent variate factorial PCA tracking method.

It is known that structural organization of PD gives function of basal ganglia circuit, but the inference is exploratory [9]. So to study the function of important component of brain is to acquire neural signal with microelectrodes [10–13]. By sampling the signals of brain during behavior, one can gain some insight into what role that part might play in behavior. Neurons within different basal ganglia nuclei have characteristic baseline discharge patterns that change with movement [9].

9.3 Methods

Twelve subjects with PD diagnosis having more than 6 years as per UK PD society brain bank criteria with good response to a precursor to dopamine cells "levodopa" and Hoehn and Yahr score of <4 with normal cognition were included in this study. The signal acquisition microrecording was performed in all patients.

The intelligent chips microelectrode implanted into the 12 PD subjects. Signal acquisition recording performed with microelectrode chips (few microns width say 10 μm) is measured at 190 Hz. MER signals were acquired with the biological amplifiers (amplified at 10,000) of the effective lead point Medtronic system by employing bootstrapping technique. They were band-pass filtered analogously with a lower cut-off frequency (0.5 kHz) and upper cut-off-frequency (5 kHz). The MER signals were sampled at 12 kHz using a 12-bit (2^N, $N = 12 = 4095$) analogue to digital conversion (A/DC) card. Later, the signals where sampled up to 24 kHz. The multiple multiunit signals were acquired for 10–20 s.

9.3.1 Latent Variate Factorial (LVF)—PCA

LVF—PCA is a well-suited technique for analyzing complex signals with interference patterns in the signals in particular brain signals. In PCA, the first eigenvector with first highest eigenvalue is the principal component 1 (PC_1).

Once eigenvectors found from covariance, the next step is to order them by eigenvalue in the order of decreasing (first PC is the highest magnitude, second PC is the second highest magnitude in the order of decreasing). Utmost at best, first three PCs are taken or considered and rest of the PCs are nullified because most of them are on the electrical baseline (zero line). PCA requires that the eigenvalues and the covariance matrix are formed. If there are n dimensions originally in the data, then n eigenvectors and n eigenvalues are calculated, and then we choose first p eigenvectors, then the final dataset has only dimensions. Considering eigenvectors, which are not nullified and forming a matrix with eigenvectors in the columns form a feature vector can be expressed as,

$$\text{Feature-vector} = [\text{Eig}_1, \text{Eig}_2, \text{Eig}_3, \ldots, \text{Eig}_n] \qquad (9.1)$$

Once a feature vector is formed, take the transpose of the vector and multiply it with the transpose matrix (vector) on the left of the original dataset matrix.

$$\text{Final data} = \text{row feature vector} \times \text{row data adjust} \qquad (9.2)$$

Here, row feature vector is a matrix with Eigen-Vectors in the column transposed such that the Eigen-Vectors are now in the rows and highest vector at the top, row data adjust is mean-adjusted data transposed, where *row feature vector* is a matrix with eigenvectors in the columns transposed so that the eigenvectors are now in the rows, with the most significant eigenvector at the top, and row data vector is the mean-adjusted data transposed.

PCA program determines mean data vector from all the row vectors within the initial data matrix fed into the program. The residual data matrix is solved by using the algorithm [1, 2, 12] Jacobi's method. Once solved, this and first three resulting PCs are stored, and the variance associated with each PC is determined and stored alongside. The purpose is to compute three PC vectors (PCVs), namely PCV_1, PCV_2, and PCV_3 of a class of signal waveforms. The program arranges the data of a single signal into a matrix of order $m \times n$ ($m < n$) by splitting the signal into *m-segments* each of *length n*.

Data minimization—This program determines PC coefficients a_1, a_2, and a_3 and calls—invokes a function to operate on the following expression [13].

$$X = G + a_1.P_1 + a_2.P_2 + a_3.P_3 + \text{error} \qquad (9.3)$$

where X = test spectral vector, G = mean of class of spectral vectors, P_1, P_2, and P_3 = first three PCs. a_1, a_2, and a_3 = PC coefficients, such that the error,

$$e^2 = \sum [X(j) - G(j) - a_1.P_1(j) - a_2.P_2(j) - a_3.P_3(j)]^2 \qquad (9.4)$$

is minimized.

Those coefficients characterize a test turning point spectrum, by a point in two or three dimensional vector space. Thus, the PC coefficients allow a representation of EMG sequence, which can be plotted in a two- or three-dimensional space (2D or 3D). The following Fig. 9.1 is obtained with MER. The subthalamic nuclei were detected by interfered noise with a bigger electrical zero line and asymmetrical discharge ejection patterns of multiple frequencies.

The STN was clearly distinguished from the dorsally located zona incerta and lenticular fasciculus (field H2) by a sudden increase in background noise level and increase in discharge rate typically characterized by rhythmic bursts of activity with a burst frequency between 20 and 35 Hz. Intra-operative recording was performed

Dorsal STNs

Target STNs

Ventral STNs

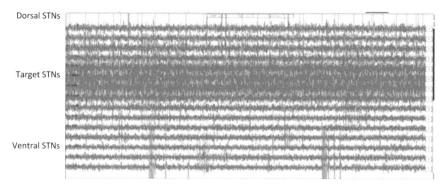

Fig. 9.1 The signal patterns of single and multiunit subthalamic nuclei neurons at range of depths

Single unit (single channel) STN

Central Channel (11 mm) typical firing pattern with irregular firing and broad baseline (from -1.00 level)

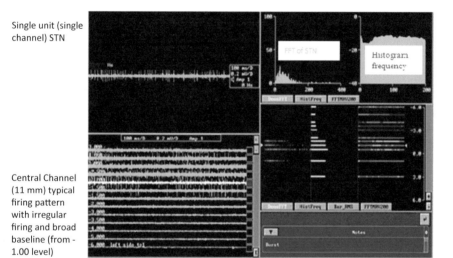

Fig. 9.2 MER signal recording. At the posterior level, STN was encountered lower than expected at that location

in all five channels. All five microelectrodes were slowly passed through the STN, and recording was performed from dimensions 10 mm above to 10 mm below the STN calculated on MRI. The STN was detected by a high noise with a large electrical baseline and an irregular discharge patterns with multiple frequencies. Figure 9.2 shows the microelectrode recording which was obtained from the STN.

Principal Component Analysis-Based Targeting Method

The STN-data centered and diffusion matrix is computed but not the STN neural Signal-data centered. The variances/10,000 for the case of the means (see Fig. 9.3a) are (**PC$_1$: 9.5056 PC$_2$: 0.9306 PC$_3$: 0.7124** 0.0466 0.0148 0.0029 0.0014 0.0005

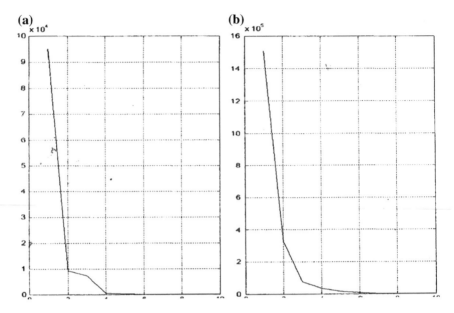

Fig. 9.3 a Variances of PC scores for means. **b** Variances of PC scores for standard deviations (SDs)

0.0002 0.0000 out of a total of 11.215). The three principal component scores are indicated with PC1, PC2, and PC3.

The variances of principal component scores for means and standard deviations are plotted in graphs (Fig. 9.3a, b). Here, the variation of the components for the SDs gives the variances of the ten component scores in the order of decreasing magnitude (1.0e+006 * 1.5079 0.3270 0.0740 0.0340 0.0151 0.0080 0.0016 0.0005 0.0001 0.0001).

Thus, out of a total variation of 1.8692×10^6, the first two components ($PC_1 = 1.5079$, $PC_2 = 0.3270$) themselves represent almost all the variations in the tables.

Thus, the first three scores account for almost all the variations observed in the data; one can effectively reproduce the data matrix by just taking the first three principal component scores for the 12 patients, and the three weight vectors (i.e., corresponding eigenvectors).

That is, the entire information about the means of the ten muscles for any patient is effectively reproduced by these three scores for that patient.

The first three principal component scores for the 12 patients are given in Table 9.1—Principal Component Analysis Vectors (**PCAV**).

The scatter plot of the first two component scores for the 12 patients is given in Fig. 9.5.

A similar analysis for the variation of the components for the standard deviations gives the variances of the ten component scores in the order of decreasing

Table 9.1 First three PC scores for the means (see Fig. 9.4a)

Group	Patient	PCAV1	PCAV2	PCAV3
D	A1	−55.86	3.39	−6.90
	A2	−60.64	5.97	−9.72
	A3	−60.05	0.79	−15.05
	A4	143.97	16.56	−0.74
	A5	144.02	16.42	−1.32
	A6	−62.56	6.64	−12.83
D	A7	146.97	15.78	−2.60
	A8	49.90	−89.75	11.39
	A9	−64.30	17.74	77.04
	A10	−64.64	4.56	−14.10
D	A11	−62.90	4.60	−13.02
D	A12	−53.88	−2.75	−12.11

PC Principal components, *D* Discordant group

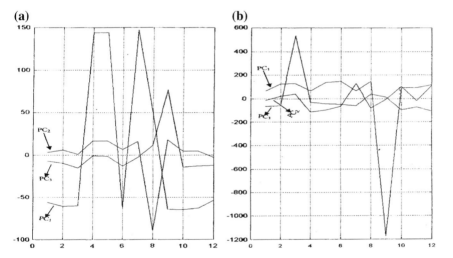

Fig. 9.4 a Variances for PCA scores for means. **b** Variances for PCA scores for SDs

magnitude, as shown before: (see Fig. 9.4b) (1.0e+006 * 1.5079 0.3270 0.0740 0.0340 0.0151 0.0080 0.0016 0.0005 0.0001 0.0001).

Thus, out of a total variation of 1.8692×10^6, the first two components themselves represent almost all the variations in the tables. The first two pc scores for the 12 patients are given in Table 9.2.

A scatter plot of these PC scores is given in Fig. 9.6.

A further attempt at clustering these 12 patients by using the principal component (PC) scores of their "differences" in counts of signals was computed, with the

Table 9.2 First 2 PC scores/ 1000 for the standard deviations (see Fig. 9.4b)

D	0.06	−0.06
	0.12	−0.06
	0.12	0.53
	0.06	−0.02
	0.13	−0.04
	0.14	−0.04
D	0.06	−0.06
	0.14	0.04
	−1.17	0.01
	0.09	−0.09
D	0.09	−0.07
D	0.11	−0.10

PC Principal Components, *D* Discordant group

Fig. 9.5 Scatter plot of principal component analysis (PCA)—first two PC scores for signal averages

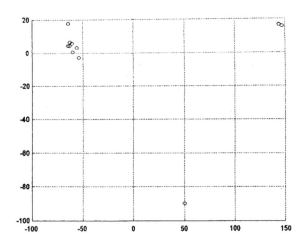

Fig. 9.6 Scatter plot of PCAs first two scores for standard deviations (SDs)

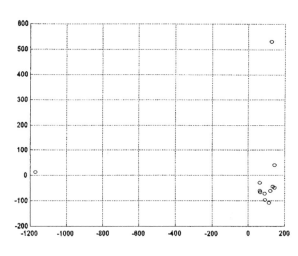

Table 9.3 Principal component scores of 12 patients

$$
Pa2 = \begin{pmatrix}
-100.7286 & 4.9914 \\
-135.4948 & 18.6261 \\
58.9665 & 31.6554 \\
-4.4131 & 0.9095 \\
-80.9136 & 1.0355 \\
-12.1537 & -18.3320 \\
7.9185 & 8.2583 \\
-49.9930 & 63.3184 \\
9.2109 & -46.9489 \\
-86.4117 & -27.6721 \\
37.5296 & 55.9556 \\
0.8076 & 82.0194
\end{pmatrix}
$$

Fig. 9.7 80% variance circa depicted in this scatter—plot

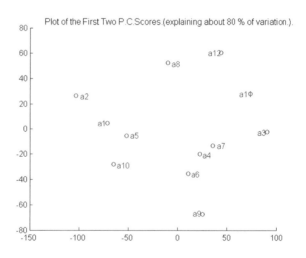

following results: The total sum of squares is 89,260, while the first two components account for $68{,}836 = 50{,}219 + 18{,}617$, i.e., accounts for about 75% of the
variation in the data. This pair of coordinates give a good enough summary representation of the information conveyed by the distance matrix. The two (centered)
principal component scores for the 12 patients are given in the Table 9.3 (see pa2).

A scatter plot of these scores is shown in the Fig. 9.7. Their distances are
summarized in the Euclidean space, which patient is nearer to and which are the
patients farthest to each other.

From this plot of the first PC scores of 12 subjects, the following conclusions are
drawn. The points are well scattered out, without clear pattern except for the case of

patients {a6, a4, a7}. These three are near enough to one another as compared to the remaining nine patients. Indeed, these three seem to form a lineal ordered set and thus forming ellipsoidal curves or resembling clouds in the space with a4 coming between a6 and a7. It is also suggestive that {a3, a11, a12}, {a10, a5} may form two similar lineal ordered sets, though the distances are much larger than in the case of the first set. Patients {a8, a9} are isolated and are very farthest to each other and thus explaining 80% variance. However, these findings are to be cross-validated with clinical findings on the same patients. It is interesting that the D group patients do not form a cluster in this scenario. This perhaps needs further looking into the clinical picture of patterns, other than the C and D groupings.

9.4 Results

Microelectrode recordings of these patients were computed by applying latent variate factor principal component analysis (PCA) method and their feature values (mean $\pm$ SD) with DBS "on" and "off." The PCs were solved for 12 subjects by using the solved eigenvectors. It is observed that the first two PCs are good enough to summarize DBS on—off states between normal and different abnormal. First eigenvector is the best mean square (BMS) fit for the feature vectors of vigorous subjects. Hence, PC1, i.e., the highest magnitude of first eigenvector, describes the amplitude of the MER signal features in relation to the mean of healthy subjects.

9.5 Conclusions

We applied PCA-based latent variate factorial analysis technique. Future study involves applying mutual information technique, coherence, and other statistical signal processing methods with controls for better understanding of Parkinson's disease with microelectrode signal acquisition of the STN.

References

1. Jankovic J (2008) Parkinson's disease: clinical features and diagnosis. J Neurol Neurosurg Psychiatry 79:368–376 [PubMed: 18344392]
2. Fahn S, Elton RL (1987) The Unified Parkinsons disease rating scale. In: Fahn S, Marsden CD, Calne DB, Goldstein M (eds) Recent developments in Parkinsons disease. Macmillan Healthcare Information, Florham Park, N.J, pp 153–163
3. Antoniades CA, Barker RA (2008) The search for biomarkers in Parkinsons disease: a critical review. Expert Rev 8(12):1841–1852
4. Morgan JC, Mehta SH, Sethi KD (2010) Biomarkers in Parkinsons disease. Curr Neurol Neurosci Rep 10:423–430 [PubMed: 20809400]

5. Strauss E, Lasker Foundation (2014) Lasker ~ DeBakey Clinical Medical Research Award to Alim Louis Benabid and Mahlon DeLong. Award description: for the development of deep brain stimulation of the subthalamic nucleus, a surgical technique that reduces tremors and restores motor function in patients with advanced Parkinson's disease, pp 1–4. http://www. laskerfoundation.org/awards/2014_c_description.htm
6. DeLong M (2014) Laskar award winner. Nat Med 20(10)
7. Benabid AL, Chabardes S, Mitrofanis J, Pollak P (2009) Deep brain stimulation of the subthalamic nucleus for the treatment of Parkinson's disease. Lancet Neurol 8:67–81 [PubMed: 19081516]
8. Squire L, Berg D, Bloom FE, du Lac S, Ghosh A, Spitzer NC (2012) Fundamental neuroscience, 4th edn. AP Academic Press
9. Andrade-Souza YM, Schwalb JM, Hamani C, Eltahawy H, Hoque T, Saint-Cyr J, Lozano AM (2008) Comparison of three methods of targeting the subthalamic nucleus for chronic stimulation in Parkinson's disease. Neurosurgery 62(Suppl 2):875–883
10. Benabid AL (2003) Deep brain stimulation for Parkinson's disease. Curr Opin Neurobiol 13 (6):696–706
11. Benabid AL et al (1994) Acute and long-term effects of subthalamic nucleus stimulation in Parkinson's disease. Stereotact Funct Neurosurg 62(1–4):76–84
12. Moran A et al (2008) Subthalamic nucleus functional organization revealed by Parkinsonian neuronal oscillations and synchrony. Brain 131(12):3395–3409
13. Defer GL (1999) Core assessment program for surgical intervention therapies in Parkinson's disease. Mov Disord 14(4):572–584

Chapter 10
CNN Data Mining Algorithm
for Detecting Credit Card Fraud

P. Ragha Vardhani, Y. Indira Priyadarshini and Y. Narasimhulu

Abstract In developing countries, credit card fraud continues to be a menace. We have to encounter this menace to emancipate humans from being the victims of credit card frauds. The most effective and powerful tool, i.e., data mining, is used by many researchers nowadays to detect and unmask the credit card frauds. Previously, many data mining algorithms were used for detecting credit card fraud. We pose a novel data mining algorithm called condensed nearest neighbor (CNN) algorithm to detect the credit card fraud. CNN algorithm is a nonparametric method used for classification. By using data reduction concept, CNN algorithm aims to form a condensed set by retaining the samples that are important in decision making.

Keywords Credit card fraud · Data mining · CNN algorithm · Nonparametric methods · Classification · Data reduction

10.1 Introduction

Fraud is nothing but "illicit or criminal diddling of others individual credit card resulting in financial gain." Credit card fraud can be a form of theft in which an individual can use others' credit card information for any kind of transactions and also can withdraw money from someone's account. Credit card fraud also includes bamboozling use of a debit card and may be attained by the theft of the factual card or by illicitly accessing the credit card owner's account and personal information, like name on card, address, card number, and card's security number.

P. R. Vardhani (✉) · Y. I. Priyadarshini · Y. Narasimhulu
Ravindra College of Engineering for Women, Kurnool, India
e-mail: ragha.vardhani@gmail.com

Y. I. Priyadarshini
e-mail: indusai.yashu@gmail.com

Y. Narasimhulu
e-mail: narasimedu@gmail.com

© The Author(s) 2019   85
N. B. Muppalaneni et al., *Soft Computing and Medical Bioinformatics*,
SpringerBriefs in Forensic and Medical Bioinformatics
https://doi.org/10.1007/978-981-13-0059-2_10

According to survey, every year, $500 million US dollars are being lost due to credit card fraud. The more we carry on online transactions, the more is the chance of we being the victims to credit card fraud.

Bolton and Hand [1] tagged two subtypes of fraud:

- Application fraud: By using false personal information, the fraudster gets the new card from the issuing companies. Then, they use all the money in short span.
- Behavioral fraud: Furtively, the fraudster knows the factual details of card, and transactions are being done on a "cardholder not present" mode.

Fraud Cycle:

- Fraud detection: Exercising detection models on several observations and communicate risk to each dimension.
- Fraud investigation: A specialist investigates whether a fraud has taken place and procures the evidence to save the victims involved.
- Fraud confirmation: With the help of field research, true fraud tag will be discovered.
- Fraud prevention: Restraining the scam not to occur in future. It results in finding the fraud before the swindler knows he will commit the swindle (Fig. 10.1).

Existing credit card fraud comes in different flavors [2]:

- Lost or stolen card: The bilker steals the legitimate card of the user and utilizes it for further transactions. The cardholder reports them missing.
- Identity Theft: By knowing the personal information of a person, we can do fraud like opening a new account on his name, other accounts.
- Skimming: It involves copying the data which exists on magnetic strip of payment card. This type of fraud is problematic to find, as victims may not be aware of fraudulent payments until their next statement arrives.
- Counterfeit card fraud: Genuine debit/credit card is being forged by the fraudster using real card magnetic stripe information.

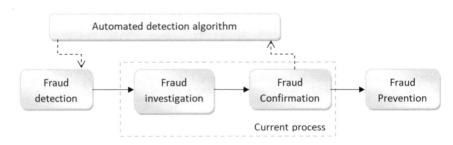

Fig. 10.1 Main activities of fraud cycle

Table 10.1 The modus
operandi of credit card frauds

Type	Percentage (%)
Lost or stolen card	48
Identity theft	15
Skimming	14
Counterfeit card	12
Phishing	6
Other	5

Table 10.2 Credit card fraud
occurrence in different
countries

Country name	Percentage (%)
Ukraine	19
Indonesia	18
Yugoslavia	18
Malaysia	6
Turkey	9
Others	30

Fig. 10.2 Pie chart for
modus operandi of credit card
frauds

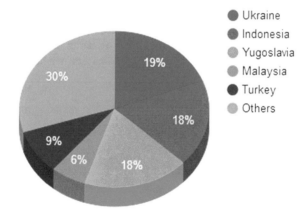

- Phishing: It is caused when someone makes you tell about your account and
  personal information and then he tricks.

   Table 10.1 reveals the modus operandi of credit card frauds. Table 10.2 [3] and
pie chart Fig. 10.2 summarize the fraud rate in various countries.

10.2   Literature Review

Shen et al. [4] applied three different models of classification like decision tree, neural networks, logistic regression to detect the different frauds occurred due to credit card transactions. They proposed a scheme to find best model that correctly finds the credit card fraud.

Esakkiraj and Chidambaram [5] proposed a model using hidden Markov model for detecting credit card fraud. This model uses some sequence of operations in online transactions to find whether the user is normal user or fraud user.

Sánchez et al. [6] used association rules to mine the knowledge and to detect the deceitful transactions done by fraudsters. In this method, they have used four levels of technological support as a part of their computerized management systems.

Wen-Fang and Wang [7] schemed a credit card fraud detection model which uses the concept of outlier detection. This method is based on the distance sum which is calculated by using infrequency and unconventionality of fraud in credit card transaction data.

Behera and Panigrahi [8] presented a method which uses three phases in detecting the credit card fraud which includes authentication phase, analysis phase, and learning phase. In authentication phase, user authentication and card details are verified. In analysis phase, by using fuzzy c-means algorithm, the normal usage patterns of genuine user are found by using their past activities. In learning phase, neural network algorithm is applied to find whether the user is genuine or fraud.

10.2.1   Data Mining

Data mining is a procedure of finding novel patterns by exploring gigantic volumes of data. Data mining is one among the most influential tools for decision support systems and plays a vital stint in credit card fraud detection. Some of the techniques used are:

  I. Decision Tree
      Decision tree is one model of classification, which is a flowchart-like structure comprising root node, internal nodes, external or leaf nodes, and branches. Root node and internal nodes represent the attributes, leaf node represents class labels, and branches represent the outcome of each attribute. Strength:

      • Extremely fast at classifying unknown records.
      • Work well in the presence of redundant attributes and able to handle noise.

Weakness:

- May suffer from over fitting.
- It has complex algorithm structure.

II. Navies Bayesian algorithm

Bayes theorem is used in Naive Bayes algorithm, which adopts independence assumption between variables. This classifier assumes the class conditional independence; that is, in a class, the values of variables are independent.

Strength:

- Easy and fast to implement.
- Require slight measure of training data to predict the parameters.

Weakness:

- There will be a chance of loss of accuracy.

III. Logistic Regression

It is a technique of regression model where target variable is having binary values: 0,1 or Y,N or T,F. Logistic regression models the data in the form of a logistic curve rather than a straight line, and approximate value of output probability is produced.

Strength:

- For data classification, it uses a simple probability formula.
- It produces fruitful outputs only when linear data is supplied as an input.

Weakness:

- Infertile results are produced when nonlinear input is supplied.

IV. Association rule mining

It is a well-known process for finding effective associations between attributes in huge databases. By using association rules, we are able to mine the frequent items from the database. Association rule mining involves two steps in process.

1. Generating frequent item set.
2. Rule generation.

Strength:

- Easy and fast to implement.
- Association algorithms produce many rules, and every rule has different conclusion.

Weakness:

- It requires many scans of database during frequent item set generation.

V. Neural Networks
   This computational model resembles functioning of neuron in human brain. Fraud detection can be done by using this concept by using the two phases, i.e., learning phase and training phase. Set of connected nodes are called neurons, which receive impulse from other nodes to perform some operation on input and transmit the result to output nodes. Neurons are layered in such a way that output from the above layer is supplied as input to the layer below it.
   Strength:

- Ability to classify untrained patterns.
- For extracting rules from neural networks, new techniques have been evolved.

Weakness:

- Difficulty to confirm the structure.
- Lack of available dataset.
- Require a number of parameters.
- Poor interpretability.

VI. K-Nearest Neighbor algorithm
   KNN is a nonparametric and lazy learner algorithm for classification. It uses local proximity (neighborhood) to classify the data. By using local proximity, it retrieves the k-nearest elements which are similar to one another.
   Strength:

- Basic problems are solved easily by using KNN.
- If the training data is large, efficient results are obtained.

Weakness:

- Large storage requirements.
- Highly affected to the curse of dimensionality.

The above methods are summarized in Table 10.3.

10.3 Existing Concept

By using the concept of nearest neighbor classification, we are going to detect the credit card fraud. By using KNN concept, we can easily find the credit card frauds.

KNN is an example of instance-based learning. The k-nearest neighbor algorithm [9] is a modest algorithm which stores all available instances. It classifies the

Table 10.3 Different credit card fraud detection techniques

S. No.	Technique	Description	Algorithms/methods used	Application areas
1.	Decision tree	It is an approach for classifying the data	ID3, CART, C4.5, etc.	Machine learning, decision support systems, credit card fraud
2.	Navies Bayesian classification	It uses class conditional independence among predictors	Bayes theorem	Medical diagnosis, weather forecast, credit card fraud
3.	Neural networks	It processes information using a connectionist approach	Back propagation algorithm	Pattern recognition, weather forecasting, credit card fraud
4.	Logistic regression	Depicts the association between the dependent and a set of independent attributes	Regression method	Statistical data analysis, credit card fraud
5.	Association rule mining	It discovers the relationship among the generated frequent items	Apriori, FP-growth, ECLAT	Market basket analysis, credit card fraud
6.	KNN algorithm	It stores and classifies new cases based on a similarity measure	Uses Euclidean distance, Manhattan distance, and Minkoswki distance	Gene expression, protein–protein interaction, credit card fraud

data by using similarity measures between the instances. For finding the similarity measure, it uses distance formulae like Euclidean distance. In this process, any incoming transaction is classified by calculating a nearest point to it. If the nearest neighbor is found deceitful, then the transaction is indicated as a fake. The disadvantages of KNN are [10–12].

- High computational complexity.
- Uses a large number of samples to work well.
- Curse of dimensionality.

10.4 Proposed Concept

The credit card transaction may contain a large number of instances, where we cannot use KNN concept for finding the fraud. The pitfall of KNN concept is unrelated, and irrelevant instances have large negative impact on the training process. This subdue is solved by proposing a new concept called condensed nearest neighbor (CNN) concept.

The basic notion of CNN algorithm is to condense the no. of training samples by eliminating the data which shows similarity and do not add extra information.

1. The first instance is copied from T to S
2. Implement the following: Increasing I by unity to the number of instances in T per epoch
a. Classify each instance $x_i \in T$ using S as a prototype set;
b. If a pattern x_i classified incorrectly then add the pattern to S, and go to step 3;
3. If i is not equal to the number of instances in T, then go to step 2;
4. Else the process aborts.

Fig. 10.3 CNN algorithm

Many analysts have discoursed the problem of reducing the size of training set. Hart [13] posed CNN algorithm for training set size reduction. CNN starts with the sample selected from the training set which forms the premier condensed set. This condensed set is used to classify the instances present in the training set. If any pattern in the training set is misclassified, then it is added in condensed set.

First iteration is done after scanning the training data once (Fig. 10.3).

The presented algorithm finds a subset S of the training set T such that every instance of T is closer to the instance of S of same class than that of different class. In this manner, the subset S can be used to classify all the instances in T correctly.

This process is continued until no misclassified patterns. The final condensed set obtains the two properties.

- Condensed set guarantees 100% accuracy over the training set.
- Condensed set is a subset of original training set.

Advantages

- Training data is minimized.
- Improves processing time and reduces memory requirements.
- Decreases the recognition rate.

10.5 Conclusion and Future Work

The exploration of survey related to credit card fraud detection using different data mining techniques presented in this paper culminated that all techniques use more number of attributes. To subdue this situation, a data mining algorithm called CNN algorithm has been propounded. It aims to reduce the no. of attributes for comparison, thus forming a condensed training set. It uses nearest neighbor concept along with reducing no. of attributes for comparison, thereby improving query time and memory requirements. The further work may try to reduce the computational complexity while forming condensed training set. The proposed algorithm can be applied for distributed data mining applications in detecting credit card frauds in future.

References

1. Bolton RJ, Hand DJ (2002) Statistical fraud detection: a review. Stat Sci 17(3):235–255
2. Report from https://financialit.net/news/security/credit-card-frauds-101-modus-operandi
3. Chaudhary K, Yadav J, Mallick B (2012) A review of fraud detection techniques: credit card. Int J Comput Appl 45(1):0975–8887
4. Shen A, Tong R, Deng Y (2013) Application of classification models on credit card fraud detection, "IEEE transactions June 2007". Int J Eng Res Technol (IJERT) 2
5. Esakkiraj S, Chidambaram S et al (2013) A predictive approach for fraud detection using hidden Markov model. Int J Eng Res Technol (IJERT) 2
6. Sánchez D, Vila M, Cerda L, Serrano J (2009) Association rules applied to credit card fraud detection. Expert Syst Appl 36:3630–3640
7. Wen-Fang YU, Wang N (2009) Research on credit card fraud detection model based on distance sum. In: IEEE International joint conference on artificial intelligence
8. Behera TK, Panigrahi S (2015) Credit card fraud detection: a hybrid approach using fuzzy clustering & neural network. In: IEEE International conference on advances in computing and communication engineering
9. Cover TM, Hart PE (1967) Nearest neighbor pattern classification. IEEE Trans Inf Theory 13 (1):21–27
10. Imandoust SB, Bolandraftar M (2013) Application of K-nearest neighbor (KNN) approach for predicting economic events: theoretical background. Int J Eng Res Appl (IJERA) 3(5)
11. National Center for Biotechnology Information. http://www.ncbi.nlm.nih.gov
12. K-nearest neighbors concept by Javier Bejar 2013. http://www.cs.upc.edu/~bejar/apren/docum/trans/03d-algind-knn-eng.pdf
13. Hart PE (1968) The condensed nearest neighbor rule. IEEE Trans Inf Theory (Corresp.) IT-14:515–516

Chapter 11
Study on Computational Intelligence Approaches and Big Data Analytics in Smart Transportation System

D. Venkata Siva Reddy and R. Vasanth Kumar Mehta

Abstract Computational Intelligence helps to answer the existent challenges of machine learning and Big Data. When we address about handling the data which is huge and variable, then comes various problems which may be single or multi-objective or static or dynamic in nature. Smart transportation is one such task where it has a lot of information to be handled and stored. It requires an efficient means to perform multiple tasks, and this should be done with utmost care and accuracy. Hence, a data mining algorithm with Hadoop MapReduce techniques can be used to perform the traffic regulation with accuracy and dynamically. There is a set of work done on executing a variety of algorithms in handling Big Data analytics and computational intelligence approaches. In this paper, we will highlight the characteristics of various algorithms so far those are capable of handling Big Data analytics.

Keywords Machine learning · Computational intelligence · Big data analytics Smart transportation

11.1 Introduction

With the growing trend in day-to-day life style and occupations, there is also need to set up new trend in transportation services. Already, there has been a lot of work done in improving or developing smart cities with smart transportation system. Still there is a lot of work to be done to improve the system. Smart transportation is a small integral part of smart cities. This small integral part itself has a vast data to be covered in order to monitor and also handle the system properly.

D. V. S. Reddy (✉) · R. V. K. Mehta
Department of Computer Science and Engineering, SCSVMV University,
Kanchipuram 631561, Tamil Nadu, India
e-mail: lionshivareddy@gmail.com; sivareddydv@btcollege.org

R. V. K. Mehta
e-mail: vasanth.mehta@kanchiuniv.ac.in

© The Author(s) 2019
N. B. Muppalaneni et al., *Soft Computing and Medical Bioinformatics*,
SpringerBriefs in Forensic and Medical Bioinformatics,
https://doi.org/10.1007/978-981-13-0059-2_11

With the vast growth in population, there is also increase in utility of various transportation networks. Thus, handling, monitoring, and also maintaining proper or effective transportation system have turned out to be biggest challenge in most of the smart cities. There are a lot of aspects to be considered when we speak about smart transportation. First and foremost thing is to collect information regarding the areas where there are heavy traffic and means to control accidents in such areas. Secondly, there is a need to monitor [1] and also identify the trespassers who have violated traffic rules/signals. Also, there should be an effective GPS system to provide required data to the passengers/citizens/strangers regarding the transportation facilities, departure and arrival timings, road maps with proper indications, etc. (Fig. 11.1).

When we start speaking about all these matters, it is clear cut that simple or conventional systems are not sufficient. Thus, for past few years, a lot of work has been done to introduce the communication technology with conventional system so that data collection, monitoring, analyzing, and also storing can be done more efficiently (Fig. 11.2).

New trends in transportation system have certain challenges to be faced like lack of connection between two different areas/concepts that are to be linked to draw certain important data, unevenness in the storage data in certain field of areas, lack of proper acquisition of data provided by sensors, etc. Although ICT introduced a lot of ad hoc

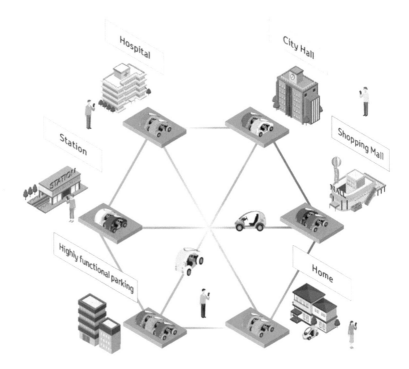

Fig. 11.1 Transportation system

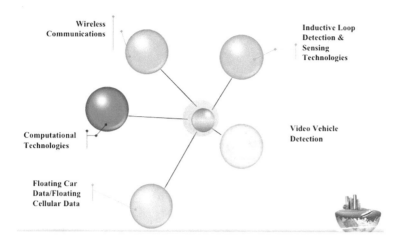

Fig. 11.2 Intelligent transport system

methods with GPS linked systems [2] to monitor transport system, still there is a problem in storing data. In order to overcome this, big data is being introduced so as to overcome the difficulties that have been faced in storage, analyzing, and managing traffic-related data. Since there is a massive data in smart transport system, big data with MapReduce model in Hadoop system can be introduced to handle and maintain the required data in transportation system (Fig. 11.3).

Since we have already emphasized the point that smart transportation involves massive data, we are intended in mainly focusing on small part of transportation system [1, 2] using machine learning techniques in data mining and big data. Since our main aim is to develop an AI system to provide and store information with regard to travelers, i.e., different transportation systems available, transportation facilities to different places from and also to the smart cities, road maps with easily identified symbols, also near and time-saving routes to reach the destiny in time.

We are intended to introduce a simple and effective algorithm in data mining and then transfer to Hadoop system [3, 4] which has the capability to maintain, handle, and also store big data. For this, we are analyzing important algorithms and their significant features with big data. In this paper, we are mainly focusing on the important algorithms that can be introduced in handling smart transportation. There are two main classes of technologies in handling big data.

11.1.1   Operational Big Data

It comprises of structure that supplies operational ability for real time, interactive workloads where data is principally captured and stored. NOSQL big data systems are designed to attain advantages of novel cloud computing architectures to contain

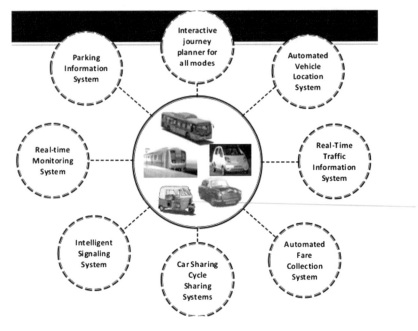

Fig. 11.3 Smart transportation system

the enormous calculations to be run economically and resourcefully. This builds operational big data workload a lot effortless to administer and to execute some of the NOSQL structures.

11.1.2 Analytical Big Data

It consists of skills like extremely analogous processing database systems and MapReduce (Hadoop) to offer logical competence for demonstration and compound investigation that may handle most or all of the data. MapReduce offers new technique of scrutinizing facts that is harmonizing to the potentials supplied by SQL and a structure supported by MapReduce that can be ranged up from single server to thousands of high- and low-end machines.

11.1.3 Importance of Hadoop

(a) Ability to store and process huge amount of any kind of data quickly.
(b) Computing power.

(c) Fault tolerance.
(d) Flexibility.
(e) Low cost.
(f) Scalability.

11.2 Related Work

We are summarizing the different research work that has been done in the field of area. What we have found is that there are certain works done and architectures have been proposed but most of them are based on IOT and GPS system. What we have found is that only a little work has been done considering data mining and big data in traffic control and monitoring. Actually, we are trying to develop an algorithm which can help in simplifying the monitoring and analyzing process in traffic control and also which can be applied to both single and multiple tasking.

1. Xia and et al. have developed Hadoop distributed platform along with MapReduce processing for data mining over real-time GPS data for different purposes.
2. Zaheer khan and et al. proposed an architecture which provides basic components to build necessary functionalities for a cloud-based big data analytical services for smart cities. They developed a prototype using MapReduce that demonstrates how cloud infrastructure can be used to analyze a sample set of Bristol open data. The prototype has been implemented using Spark and Hadoop. Finally, they compared the result and finally concluded that Spark is much faster and appropriate.

11.3 Algorithms

The algorithm should be able to handle the dynamically changing data and to adjust the target of data analytics. Handling large-scale data with a good performance in limited time should be concerned in the big data analytics. It is a great challenge to deal big data. The datasets are characterized in terms of huge volume in quantity, high variety in type or classification, velocity in terms of real-time requirements, and constant change in data structure or user interpretations. Basically, it is very tedious task to understand the data. The challenges include data analysis, data access, capture, sharing, storage, transfer. Therefore, computational intelligence is required to solve real-world problems. Some of the computational intelligence techniques are swarm intelligence, neural networks, k-means, genetic algorithms, rough set theory, etc. Of many techniques here, we are mainly focusing on important algorithms, namely swarm intelligence, k-means cluster.

11.3.1   Swarm Intelligence

Swarm intelligence is a relatively new subfield of computational intelligence (CI) which studied the collective intelligence in a group of simple individuals. With the swarm intelligence method, valuable records can be gained from the contest and collaboration of folks. On average at hand, two types of approaches are there to convey swarm intelligence as data mining techniques [5]. The first class includes procedures where folks of a swarm shift through a solution space and explore for solutions to the data mining task. There is a search approach. The swarm intelligence [6] is useful to optimize the data mining techniques. In second group, swarms shift data occurrence that are located on a low-dimensional mapping solution of the data. This is data organizing approach. The swarm intelligence is directly applied to the data sample. Swarm intelligence is mainly particle swarm optimization or ant colony optimization algorithms which is used in data mining to resolve single goal and multiobjective troubles.

Big data involves high-dimensional problems and a large amount of data. Swarm intelligence studies the collective behaviors in a group of individuals. It has shown significant achievements in solving large-scale, dynamical, and multiobjective problems. With the application of the swarm intelligence, more rapid and effective methods can be designed to solve big data analytics problems.

11.3.2   K-Means Clustering

k-means clustering algorithm is a classical algorithm based on splitting method. Since the theory of the algorithm is reliable, simple, and convergent rapidly, k-means algorithm is widely used. Even though the traditional clustering mining optimization algorithm has good accuracy in the face of massive data, its time complexity of serial calculation method is high. Aiming at the defects of traditional algorithm, there are some articles which proposed and improved method for selecting initial clustering centers and put forward a k-means algorithm optimization based on Hadoop cloud computing platform. The optimization algorithm improves the efficiency of algorithm, and the accuracy is also enhanced.

The dataset processed by MapReduce should have such characteristics. It can be broken down into many small datasets, and each small dataset can be completely parallel process. The process of k-means algorithm based on Hadoop mainly has two parts, the first part is to initial clustering centers and divide the sample dataset into a certain size of data blocks for parallel processing. The second part is to start the Map and Reduce task for parallel processing of algorithm in time, until process gets the clustering result.

Advantages of *k*-means clustering are:

(a) Fast, robust, and easier to understand.
(b) Relatively efficient.
(c) Gives best result when datasets are distinct or well separated from each other.

11.4   Discussion

Finding a simple and effective algorithm is first and foremost important task in succeeding job requirements. Here, since we are concentrating on smart transportation, we have tried to focus on simple and easily accessible algorithm to handle one or more tasks at a time. Based on the features of the two algorithms which we have highlighted, we are planning to create an effective algorithm using ant colony optimization technique which is based on Hadoop (MapReduce) platform. Smart transportation is very difficult and heavy task to handle because there are many tasks to be handled at a time. Hence, the algorithm with big data analytics should be in such a way that it can perform multitasking and also can be able to take proper decisions when needed. For instance, when we consider very busy traffic area, then big data technology should be able to monitor traffic violations if any and simultaneously should be able to pass the information to the right persons or should send alerts to the vehicle owners regarding the traffic conditions or precautions to be taken in case of bad weather or accidents. Since we are considering challenges with regard to single and also multiple tasking, we are interested to work out efficient algorithm using swarm optimization with Hadoop technology.

11.5   Conclusion

Smart transportation system can be managed efficiently and effectively with big data technology. Since big data techniques have the capacity to maintain, store, and analyze huge information, it will be possible to prevent unwanted incidents, handle traffic properly, and also prevent or even reduce losses occur due to accidents or other calamities.

References

1. Amini S, Gerostathopoulos I, Prehofer C (2017) Big data analytics for real-time traffic control. IEEE, p 10
2. Chen W et al (2015) Big data for social transportation. IEEE, p 6
3. Gadekar H, Bhosale S, Devendra P (2014) A review paper on big data and hadoop

4. Minnelli M, Dhiraj A, Chambers M (2013) Big data and big data analytics. s.l.: Willey and Sons Inc
5. Yu T et al (2015) Swarm intelligence optimization algorithms and their application. AISeL, pp 200–207
6. Cheng S et al (2013) Swarm intelligence in big data analytics. Springer, Berlin, pp 417–426
7. Mitchel T (1997) Machine learning . s.l.: McGraw-Hill
8. Rittinghouse JW, Ransome JF (2009) Cloud computing, a practical approach. s.l.: McGraw-Hill Osborne Media
9. Khan S, Shakil KA, Alam M (2015) Cloud-based big data analytics—a survey of current research and future directions. arXiv

Chapter 12
A Survey on Efficient Data Deduplication in Data Analytics

Ch. Prathima and L. S. S. Reddy

Abstract Nowadays, the demand of data safekeeping capacity is increasing dramatically. Because of more requirements of safekeeping, the computer world is appealing to toward cloud safekeeping. Security of data and cost factors are essential issues in cloud safekeeping. A duplicate document not only waste storage, it also escalates the access time. Therefore, the recognition and removal of duplicate data can be an essential task. Data deduplication, a competent method of data decrease, has gained increasing attention and recognition in large-scale storage space systems. It minimizes redundant data at the data file or subfile level and recognizes duplicated content by its cryptographically secure hash signature. It is very complicated because neither duplicate data do not have a standard key nor they contain mistake. Within this paper, the backdrop and key top features of data deduplication is preserved, then summarize and classify the data deduplication process in line with the key workflow.

Keywords Deduplication · Chunking · Hashing · CDC · Encryption

12.1 Introduction

With the tremendous development of digital data, the necessity of storage area is consistently increasing. As the quantity of digital information is increasing, the storage area and copy of such information is now an extremely challenging task. The brand new kind of storage area that is attaining much attention in current circumstance is cloud storage area. On one aspect, the digital data is increasing; on other aspect, backup problem and devastation recovery have become critical for the info centers. The three 1/4 of digital information is redundant by a written report of

Ch.Prathima (✉) · L. S. S. Reddy
K L University, Vaddeswaram, Guntur, Andhra Pradesh, India

Ch.Prathima
Data Analytics Research Lab, Department of IT, Sree Vidyanikethan Engineering College,
Tirupati, Andhra Pradesh, India

© The Author(s) 2019
N. B. Muppalaneni et al., *Soft Computing and Medical Bioinformatics*,
SpringerBriefs in Forensic and Medical Bioinformatics,
https://doi.org/10.1007/978-981-13-0059-2_12

Microsoft research. The idea of deduplication is a strategy to prevent the storage space of redundant data in storage area devices. Data deduplication is attaining much attention by the research workers because it is an effective method of data decrease. Deduplication recognizes duplicate items at chunk level by using hash functions and eliminates redundant items at chunk level.

The two important aspects to evaluate the deduplication system are: (i) the deduplication ratio and (ii) performance.

Deduplication can be carried out at chunk level or document level. Chunk-level deduplication is recommended over document-level deduplication since it recognizes and eliminates redundancy at a finer granularity. Chunking is one of the key to increase ratio and performance. The chunking and the chunk fingerprinting will be the steps that are CPU intense, whereas fingerprint indexing and querying as well as data storing and management step need a lot of ram and drive resources. He et al. increase the performance of fingerprint indexing [1].

In this study, Sect. 12.2 addresses the Deduplication Tool. In Sect. 12.3, data Deduplication is shown in details. Section 12.4 discusses Deduplication Storage area System. In Sect. 12.5, Encryption in Deduplication and Chunking Techniques are mentioned. Section 12.6 addresses the Hash System, Sect. 12.7 presents Routing Strategies. In Sect. 12.8, describes it's Advantages and is concluded with Summary in Sect. 12.9.

12.2 Deduplication Tool

The term Big data is presented because the development of data is extreme and its handling cannot be achieved using traditional data handling applications. The top level of data—both organized and unstructured is referred to as Big data [2]. For instance, vast amounts of photographs are daily distributed on face book from its users. Nowadays, the avail of data is increasing daily.

Hadoop
Hadoop [3] is an available source Java-based coding framework. It was created by computer scientist Doug Cutting and Mike Cofarella in 2006. It really appeared as a foundation for big data processing responsibilities such as scientific analytics, business planning. It offers the parallel processing capacity to handle big data. Hadoop kernel provides the framework of essential your local library required by other Hadoop module. Hadoop Distributed Data file System (HDFS) is competent of storing data across 1000s of servers. It was developed using distributed data file system. It holds substantial amount of data sent out across multiple machines. These types of files are stored in redundant fashion across multiple systems in order to avoid system from data losses. Hadoop yet Another Resource Negotiator (YARN) provides resource management and scheduling for user applications. Hadoop MapReduce is a YARN-based system for parallel processing of large datasets.

12.3 Data Deduplication

Data deduplication is the process of eliminating the repetitive data [4]. This is the technique that is employed to track and eliminate the same chunks in a storage unit. This is an efficient way to store data or information. The process of deduplication is shown in Fig. 12.1. Deduplication is generally divided into three parts, i.e., data unit based, location centered, and disk placement centered. The classification of deduplication is shown in Fig. 12.2.

Data deduplication strategies are classified into two parts, i.e., file-level deduplication and block chunk-level deduplication. In file-level deduplication, if the hash value for two files is similar then they are considered identical. Just one copy of a file is stored. Duplicate or redundant replications of the same data file are removed. This type of deduplication is also known as single-value storage (SVS). In wedge (chunk)-level deduplication, the file is fragmented into blocks (chunks) then inspections and removes duplicate obstructions within files. Only one copy of every block is maintained. It reduces more space than SVS. This may be further divided into fixed-length and variable-length deduplication. In fixed-length deduplication, the size of each wedge is regular or set, whereas in variable size deduplication, the size of each block is different or not fixed. In location-based deduplication, process

Fig. 12.1 The deduplication process

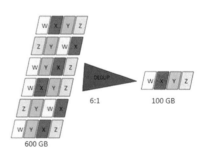

Fig. 12.2 Classification of deduplication process

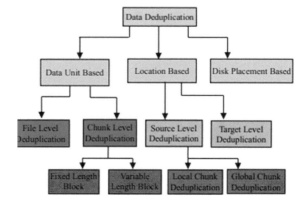

is performed on different location. It is further divided into two parts, i.e., resource-based duplication and goal-based deduplication. Before the transmission of actual data to the target, deduplication is performed. It reduces the huge amount of the backup data that might be sent through network.

With this process, duplicate data is determined before the transmitting within the network. It can be further divided into two parts, i.e., local chunk deduplication and global chunk deduplication. In the local chunk level, redundant data is removed before sending it to the destination. Inside the global chunk level, repetitive data is removed at global site for each and every consumer.

In target level, deduplication process is performed through mobile agents at the backup storage. The mobile agent tracks the redundancy at backup server and then only the unique data blocks are stored to hard disks.

Disk placement-based deduplication is based how data is to be located on disk. Forward reference point or backward references are used. Forwards reference, recent data portions are maintained and all the old data portions are associated to the recent chunks through hints. Backward reference introduces a lot more fragmentation for the earlier data chunks.

A. *Info Deduplication-Performance Evaluator* [4]

The efficiency of any data deduplication application measures:

1. Dedupe ratio where dedupe rate = scale actual data/scale data after deduplication.
2. Throughput (MBytes of DD/s).

B. *Algorithm/Working Technique* [4]

The significant strategy of deduplication process is shown in Fig. 12.3. The subsequent steps are being used for this process.

1. Divide the input data into portions or blocks.
2. Hash value for each and every block needs to be calculated.
3. The values that are made can be used to check whether the blocks of same data exists in another stored block data.
4. If perhaps duplicate data is found, then the reference to be created in data bases.
5. Based on the results, the duplicate data is eliminated. Only a unique chunk is stored.

Fig. 12.3 Dedulpication workflow

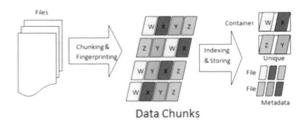

12.4 Deduplication Storage System

There are many deduplication storage space systems that are preferred for different storage area purposes.

HYDRAstor [5] is highly scalable secondary storage solution. It is predicated on decentralized hash index for the grid of storage space nodes at the backend and traditional document interface at front end. It successfully organizes large values, adjustable size and content addressed, highly resilient data blocks with the aid of directed acyclic graph.

MAD2 [6] is famous and trusted for its accuracy and reliability. It provides a precise deduplication backup service which mostly works on both at the document level with the crunch level. Technique are hash bucket matrix, a bloom filter array, a distributed hash table-based load balancing and dual cache—to be able to attain the desired performance.

Duplicate Data Eradication (DDE) [7] exactly calculates the matching hash values of the info blocks prior to the actual value of data at client side. It works on the combination of copy-on-write, sluggish improvements, and content hashing to recognize and coalesce similar blocks of data in SAN system.

12.5 Encryption in Deduplication

Deduplication works by removing the redundant blocks, files, or data. In whole data file hashing, a complete file is first sent to hashing function. The preferred hashing functions are MD5 and SHA-1. Its result is a cryptographic hash which forms the basis for the identification of complete duplicate file. The complete file hashing has fast execution with low calculation and low metadata expense. But it prevents from matching two files that just differ by a byte of information divides two types, fixed-size chunking and variable-length chunking, can be used for dividing a file. In fixed size chunking, a file is divided into an amount of static or set sized pieces called "chunks," whereas in variable size chunking, data is broken down into chunks of differing length. The cryptographic hash function is applied to the broken bits of data file to calculate the "chunk designator." The chunk designator is the key Variable in locating redundant data.

Chunking-based Data Deduplication [8]
In this method, each file is firstly divided into a sequence of contradictory pads, called chunks. Each portion is a contiguous pattern of bytes from the file which is refined independently.

1. *Permanent Chunking* [9]

The permanent or fixed size chunking mechanism divides the file into same size chunks. Figure 12.4 shows the working of fixed size chunking. This also causes a serious problem called "boundary shift." Boundary shifting problem occurs due to modifications in the information.

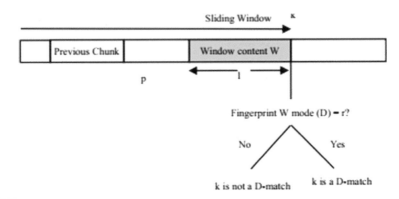

Fig. 12.4 Basic moving window

2. Content-Defined Chunking [9]

To overcome the challenge of boundary shift, content-defined chunking (CDC) is used. CDC reduces the amount of duplicate data found by the data deduplication systems, rather than setting limitations at multiples of the limited chunk size. CDC approach defines breakpoints where a specific condition becomes true.

3. Basic Moving Window (BMW) [10]

BMW uses the model of the hash-breaking (non-overlap) approach. The three main parameters needed to be pre-configured are: a predetermined size of window W, an integer divisor—D, and an integer remainder—R, where $R < D$. In BMW method, in each shifts in the window, fingerprint will be made and check for the chunk boundary. This is shown in Fig. 12.4.

4. Improved Moving Window (IMW) [11]

It includes the good thing about fixed portion size k byte. In case the fingerprints of new portion have same hash value, then consider it as redundancy and move the window forward over this data chunk; otherwise, predictions are made that the chunk may be redundant that has been changed by insertion and deletion. This is shown in Fig. 12.5.

5. Byte Index Chunking [12, 13]

Byte index chunking is source-structured deduplication network data file system. In source-structured network data file system, deduplication is conducted on client area. Index table is exchanged between source and server before non-deduplicated blocks are moved for storage area.

6. Multilevel Byte Index Chunking [14, 15]

When the chunks of large size are placed, then it will consume less amount of time in handling but deduplication rate lowers and vice versa. In multilevel byte

Fig. 12.5 Improved moving
window

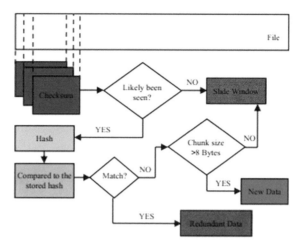

index chunking, complete file is split into multiple size of chunks in line with the
size of data file. If the quality is leaner than 5 GB, then document deduplication
process is performed using 32 KB size index table.

7. *Frequency-Structured Chunking Algorithm* [16]

FSC strategy is a cross-chunking algorithm that divides the byte stream in line
with the chunk rate of recurrence. First, it recognizes the set size chunks with high
rate of occurrence using bloom filtration then your appearance is inspected for
every single chunk in the bloom filtration.

8. *Improved Frequency-Structured Chunking* [17]

It overcomes the disadvantage of FSC algorithm. With this, CDC algorithm can
be used to chop the byte stream into chunks. This technique contributes to upsurge
in amount of metadata and it generally does not allow re-chunking when the chunk
size is set.

9. *Bimodal Content-Defined Chunking (BCDC)* [18]

A major drawback of CDC is usually that the chunks of really small size or large
size are produced. The chunks of small size cause more CPU over head and large
size causes low deduplication rate. To beat this issue, bimodal CDC algorithm is
launched. It really is one of the correct chunking algorithms that make a decision
whether to choose the chunks of really small size or large size.

10. *Multimodal Content-Defined Chunking (MCDC)*

In MCDC, the chunk's size is set regarding to compressibility information.
Large data files with low compressibility are hardly altered for large chunks and
data files with high compressibility end result into chunks of small size. With this to
begin with, the data items are split into the set size chunks then your compression
percentage is approximated of set size chunks.

11. *Fast CDC* [19]

This approach removes the chunks of really small size. It really is 10 times faster than Rabin-structured CDC and three times faster than AE-structured approach. Rather than using the Rabin fingerprint hash device, it creates the hash value for every single byte. It normalizes the chunk size circulation to small given region. The main element idea behind the performance of fast CDC is use of three techniques: (1) Optimizing hash judgment, (2) Subminimum chunk cutpoint missing, and (3) Normalized Chunking.

12.6 Hashing Mechanism for Deduplication

Hash collisions are potential problem with deduplication. The hash amount for each and every chunk is made using an algorithm. Some frequently used hashing algorithms are Rabin's algorithm, Alder-32, Secure Hash Algorithm-1 (SHA-1), and Message digest (MD-5) algorithm. Advanced Encryption Standard (AES) is hardly used for this function. The encrypted data is out-sourced to the cloud environment. Rabin's fingerprinting structure generates fingerprints using polynomials. It generates cryptographic hash of every stop in a data file.

12.7 Routing

Cluster deduplication mainly contain—backup consumer, metadata management, deduplication server node. Two types of routing systems which are being used to route the info packet, i.e., stateless routing device and stateful routing device. In stateless routing, the hash table is utilized to look for the way to server node. Stateful routing requires the info of deduplication server node to course the packet to the server node that gets the same packet.

A. *AR Dedupe* [20]

To overcome the challenge of weight balancing and communication over head, AR dedupe strategy is developed. In AR dedupe, routing server can be used. Hash table including hash information and Id ofserver node of the super chunk is stored on routing server then super chunk is routed corresponding to this identification. AR dedupe mainly contains four parts—backup consumer, metadata server, deduplication server node, and routing server, shown in Fig. 12.6.

B. *Boafft* [21]

Boaftt uses the hash function or Jaccard coefficient to gauge the similarity between datasets. Client divides the info stream into smaller chunks. Fingerprints are manufactured for each and every chunk independently and coordinate these chunks into excellent blocks. Boaftt locates the routing address for these chunks as the feature fingerprint by getting together with metadata server and route the

Fig. 12.6 ARDedup

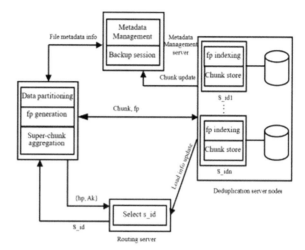

superblock to appropriate data server. Fingerprints are stored on data server. Boafft uses the box to store fingerprint and coordinate the fingerprint in similarity index then boafft compare the superblock with similarity index. The working of Boafft is shown in Fig. 12.7.

12.8   Data Deduplication: Advantages

Data deduplication optimizes the storage area and will save a lot of money by minimizing the storage area and bandwidth cost. This system needs less hardware to store the info because the storage area efficiency is increased that reduces the hardware cost too. It reduces the backup costs because buying and preserving less storage area will return us with the faster results. If effectively escalates the network bandwidth and boosts the network efficiency.

12.9   Conclusion

This paper mainly targets the data deduplication, its chunking techniques, and storage area systems. The primary goal of data deduplication is to eliminate redundant data and keep maintaining unique backup of documents. Data deduplication can be a trend in market and good alternative of data compression. It really is one of the smart storage saving device. Data deduplication well handles the storage areas and boosts the network bandwidth which is necessary for moving data.

Fig. 12.7 Boafft

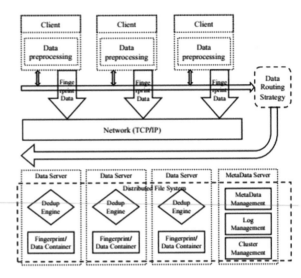

References

1. He S, Zhang C, Hao P (2009) Comparative study of features for fingerprint indexing. In: 16th IEEE international conference on image processing (ICIP), Cairo
2. Fang H, Zhang Z, Wang CJ, Daneshmand M, Wang C, Wang H (2015) A survey of big data research. IEEE Netw 29(5):6–9
3. Panda M, Sethy R (2015) Big data analysis using Hadoop: a survey. Int J Adv Res Comput Sci Softw Eng 5(7):1153–1157
4. Malhotra J, Bakal J (2015) A survey and comparative study of data deduplication techniques. In: International conference on pervasive computing (ICPC), Pune
5. Dubnicki C, Gryz L, Heldt L, Kaczmarczyk M, Kilian W, Strzelczak P, Szczepkowski J, Ungureanu C, Welnicki M (2009) HYDRAstor: a scalable secondary storage. In: 7th USENIX conference on file and storage technologies (FAST 09), San Francisco, California
6. Wei J, Jiang H, Zhou K, Feng D (2010) MAD2: a scalable high-throughput exact deduplication approach for network backup services. In: IEEE 26th symposium on mass storage systems and technologies (MSST), Incline Village, NV
7. Hong B (2004) Duplicate data elimination in a SAN file system. In: 21st international conference on massive storage systems and technologies (MSST), College Park, MD
8. Xia W, Jiang H, Feng D, Douglis F, Shilane P, Hua Y, Fu M, Zhang Y, Zhou Y (2016) A comprehensive study of the past, present, and future of data deduplication. Proc IEEE 104 (9):1681–1710
9. Li A, Jiwu S, Mingqiang L (2010) Data deduplication techniques. J Softw 2(9):916–929
10. Hsu W, Ong S, System and method for dividing data into predominantly fixed-sized chunks so that duplicate data chunks may be identified. US Patent US7281006B2, 2007
11. Bo C, Li ZF, Can W (2012) Research on chunking algorithms of data de-duplication. In: International conference on communication, electronics and automation engineering, Berlin, Heidelberg
12. Lkhagvasuren I, So J, Lee J, Ko Y (2014) Multi-level byte index chunking mechanism for file synchronization. Int J Softw Eng Appl 8(3):339–350

13. Lu G, Jin Y, Du D (2010) Frequency based chunking for data de-duplication. In: IEEE international symposium on modeling, analysis and simulation of computer and telecommunication systems (MASCOTS), Miami Beach, FL
14. Zhang Y, Wang W, Yin T, Yuan J (2013) A novel frequency based chunking for data deduplication. Appl Mech Mater 278:2048–2053
15. Kruus E, Ungureanu C, Dubnicki C (2010) Bimodal content defined chunking for backup streams. In: 8th usenix conference on file and storage technologies (FAST-10), San Jose, California
16. Wei J, Zhu J, Li Y (2014) Multimodal content defined chunking for data deduplication, Huawei Technologies
17. Zhang Y, Feng D, Jiang H, Xia W, Fu M, Huang F, Zhou Y (2017) A fast asymmetric extremum content defined chunking algorithm for data deduplication in backup storage systems. IEEE Trans Comput 66(2):199–211
18. Xia W, Zhou Y, Jiang H, Feng D, Hua Y, Hu Y, Liu Q, Zhang Y (2016) FastCDC: a fast and efficient content-defined chunking approach for data deduplication. In: 2016 USENIX conference on usenix annual technical conference, Berkeley, CA, USA
19. Paulo J, Pereira J (2014) A survey and classification of storage deduplication systems. ACM Comput Surv (CSUR), 47(1)
20. Xing Y, Xiao N, Liu F, Sun Z, He W (2015) AR-dedupe: an efficient deduplication approach for cluster deduplication system. J Shanghai Jiaotong Univ (Sci), 76–81
21. Luo S, Zhang G, Wu C, Khan S, Li K (2015) Boafft: distributed deduplication for big data storage in the cloud. IEEE Trans Cloud Comput 99:1–13

Chapter 13
Classification of Alzheimer's Disease by Using FTD Tree

Bala Brahmeswar Kadaru, Allada Apparna, M. Uma Maheswara Rao
and G. Trinesh Sagar Reddy

Abstract Due to the increasing demand on Alzheimer's, the continuous monitoring of health and characteristics is significant for maximizing the yields. Even though many physicochemical parameters are available for monitoring the Alzheimer's, the knowledge of domain experts is expected to analyze these parameters to find the final decision about the Alzheimer's disease. In order to utilize the knowledge of the domain experts for Alzheimer's disease, we have developed a functional tangent decision tree algorithm which predict the disease based on the physiochemical parameters. The proposed method of predicting the disease consists of three important steps such as uncertainty handling, feature selection using reduce and core analysis, classification using the functional tangent decision tree. The proposed functional tangent decision tree is constructed by utilizing a function called functional tangent entropy for the selection of attributes and split points.

Keywords Alzheimer's disease · Decision tree · Entropy · Accuracy

13.1 Introduction

Alzheimer's one of the dementia causes the problems for memory loss and thinking. It is a neuro deteriorating type of dementia which starts gently and later it turns to severe. Alzheimer's is mainly affected to the people who are having the age

B. B. Kadaru · A. Apparna · M. U. M. Rao (✉) · G. T. S. Reddy
CSE Department, Gudlavalleru Engineering College, Gudlavalleru, Andhra Pradesh, India
e-mail: umalu537@gmail.com

B. B. Kadaru
e-mail: balukadaru2@gmail.com

A. Apparna
e-mail: alladaaparna9@gmail.com

G. T. S. Reddy
e-mail: trnineshsagar93@gmail.com

© The Author(s) 2019   115
N. B. Muppalaneni et al., *Soft Computing and Medical Bioinformatics*,
SpringerBriefs in Forensic and Medical Bioinformatics,
https://doi.org/10.1007/978-981-13-0059-2_13

more than 65. But in some cases, it may also occur for the people below 65 which is called as younger-onset Alzheimer's disease (also called as early-onset Alzheimer's). In starting stages, memory loss is moderate (short-term memory loss). At last stage, it can cause problems with language, loss of impulse, and behavioral issues.

Alzheimer's and the Brain
There are some microscopic changes that takes place in brain which begins before memory loss first sign. The brain is having a 100 billion nerve cells, where each nerve cell connects to many neurons which communicates. Each of them are having a different type of tasks such as study, memorializing, deliberating. For doing their work, the neurons work like generating energy, constructing the equipment.

Scientists believe that Alzheimer's is not only the cause of memory loss. In that case, many people are having trouble with memory—this does not mean that all of them are having Alzheimer's.

The Role of Plaques and Tangles
We are having two peculiar structures called plaques and tangles which are the main suspects in demolishing and destroying the nerve cells. Plaques are having a protein particles called as "beta-amyloid" which builds the space between the nerve cells. The tangles are the twisted fibers of other protein called "tau" which builds up the inside cells. Scientists also do not know about the role played by plaques and tangles in Alzheimer's. It is the eradication of nerve cells which causes memory failure, personality changes (Fig. 13.1).

Symptoms
(1) Memory loss, (2) Dilemma in Planning and Problem Solving, (3) Facing challenges in daily tasks, (4) Places are confusing, (5) Changes in eyesight, (6) Vocabulary becomes hard, (7) Misplacing the things, (8) Making decisions lately, (9) Less involvement in work, (10) Changing the mood.

The initial stage of the Alzheimer's disease is crucial for determining. Lifetime of the people with Alzheimer's is very less and it may range from three to ten years. Rarely 3% of people live more than fourteen years.

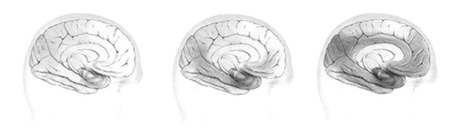

Fig. 13.1 Plaques and tangles tend to spread through the cortex as Alzheimer's progresses

13.2 Related Work

Tests to Diagnose Alzheimer's
While conducting the tests, we will be getting 90% positive anticipating value. We are having different tools for diagnosis of Alzheimer's disease.

Patient Antiquity
In this, we will be having the patient history which helps the doctor for estimating the current health of the patient. It also helps doctor for analyze and to evaluate the patient health.

Physical Tests
The physical tests are about for checking the physical condition of the patient. The physical tests include height and weight, blood pressure, pulse, bones and muscles, nerves, chest.

Laboratory Tests
The laboratory tests are to collect the samples from the body as blood test, urine test. These tests are useful to identify the diseases. In blood test, we look for the count of platelets, red blood cells, white blood cells. By these tests, the doctor can find out the symptoms of patient which are similar to that of Alzheimer's disease.

Spinal Tap
The spinal tap is also called as lumbar puncture. This is used to examine the fluid (CSF) around the spinal cord which is taken by a needle to test in a laboratory. This test is conducted for identifying the nerves condition in the nervous system.

Computed Tomography Scan
The CT scan is used for taking the number of X-rays of the patient body from different places in a short span of time. Then, the X-ray pictures are loaded into a computer which generates a order of the patient body parts. In this scan, we can identify the modifications in the parts such as contraction in size of the brain (atrophy), increasing the size of the fluid-filled chambers.

Magnetic Resonance Imaging
MRI is used to get the clear imaging of the total patient body without the use of X-rays. The MRI uses the radio waves, huge magnet, and a computer connected to it for getting the pictures. It is used to identify the dementia and also helps in identifying the changes in nerves of the brain.

13.3 Sample Data [1]

The decision trees are looking simple but they are very important role in classification. The decision trees are to be developed from the algorithms which divides the data into branches. These branches will be forming the decision tree where the

Table 13.1 Alzheimer's disease sample data

RID	A	IB	MD	G	AD
1	65–85	Yes	Yes	Male	Yes
2	65–85	Yes	Yes	Female	Yes
3	<65	No	Yes	Female	Yes
4	65–85	Yes	No	Female	Yes
5	>85	No	No	Female	Yes
6	65–85	Yes	Yes	Female	Yes

root node placed at the top of the tree. We are having different types of algorithms such as the ID3, C4.5, and classification and regression tree (CART).

In our sample data, we are having only five attributes such as gender (G), age (A), causes of genetic (CG), injuries in brain (IB), and muscular disease (MD).

In the sample dataset, we are using 20 tuples having different values varying on different patients (Table 13.1).

13.4   Data Mining Methods

ID13 [2]

The person who developed the ID3 algorithm was J. Ross Quinlan at the University of Sydney. He firstly conferred about ID3 in a book called Machine Learning in 1975. ID3 is purely based upon the Concept Learning System (CLS) Algorithm.

ID3 is improved based upon the adding a feature selection probing. ID3 searches the different kind of attributes and then it will be checking the different types of attributes, and if the attribute perfectly analyzes the datasets, then the ID3 quits. If the attribute is not about to analyze the datasets, then the ID3 recursively functions on the n (number of possible values for an attribute) and dividing into different subsets for getting their best attribute.

C13.5 [3]

This algorithm is said to be as the enhanced version of ID3; mainly in this, we are using the gain ratio as the splitting area, but in the ID3 algorithm we take gain for the splitting area in a tree. This C4.5 will be manipulating the both continuous and the discrete attributes. For manipulating the continuous attributes, we will be creating the threshold and then separating the attributes whose value is more than the threshold value and also the equal and below values. The data is to be sorted at each and every node for getting the best attribute. The attributes will perform splitting when the attributes splitted below the threshold value. The major convenience for using the C4.5 can manage the datasets having the different types of patterns and attribute values.

AFI-ID3 [4]
It is an analyzing function (AF) representing the alliance between the all elements and their attributes. In this, we are proposing the new type of heuristic function called AFI function and we also used the attribute selection method which is mainly based up on the algorithm (AFI-ID3). For the applicability for the attributes, it depends upon the values that attributes are having the maximum values such as gain, that is, selected based up on the attributes in the ID3 algorithm.

Classification and Regression Tree (CART) [5]
The CART technique is proposed by Breiman in 1984. It is mainly used for the classifications and the regression trees. In this CART, we will be splitting the attributes in construction of the classification trees. It also uses the Hunt's model for constructing the decision tree for implementing serially. By having the training dataset we perform the Pruning in the CART. It also uses the numeric and the categorical attributes for building the decision trees. CART is said to be unique from the different algorithms based upon it is using the regression analysis from the regression trees. It uses the many individual single variable splitting like gini index, symgini. Salford Systems developed an extended version of the CART called CART ® by using Breiman code. CART ® is having the upgraded features for upcoming the small faults in the CART and constructing the decision tree with the high level of classification and the prediction veracity.

13.5   FTD Tree Algorithm for Classification of Alzheimer's Data

The FTD Tree is a new version of traditional decision tree technique. This technique is used for regression and classification. It is the type of nonparametric supervised learning method. The major advantage of decision tree over the other learning algorithm is simple to check the training datasets and easy to interpret and also the cost is minimum. In the decision tree, the splitting is done by the efficient algorithm called Hunt's algorithm. In the algorithm, the Gini value is calculated for each and every node such as parent and child node. Depending upon the Gini value, the split is selected, and the minimum gini value is preferable to split. The proposed system work of FTD Tree algorithm contains two major steps such as (i) FTD Tree construction and (ii) performing classification through a FTD Tree. The definitions of construction and classification of the FTD Tree are given below.

Definition 13 The functional tangent entropy is given as follows

$$\mathrm{FE}(a_j) = - \sum_{j=i}^{u(a_j)} \mathrm{prob}^j f(\mathrm{prob}^j) \tag{13.1}$$

$$f\left(\text{prob}^{j}\right) = \frac{1}{2}\left[\log\left(\text{prob}^{j}\right) + \frac{-1}{a\tanh\left(\text{prob}^{j}\right)}\right] \qquad (13.2)$$

where $\text{FE}(a_j)$ is the functional tangent entropy of a_j, $u(a_j)$ is the number of unique values in the attributes, $\log(\cdot)$ is the logarithmic function, and $a\tanh(\cdot)$ is the inverse hyperbolic tangent function.

Definition 13 The functional information gain $\text{FIG}(a_j, a_b)$ is calculated as follows

$$\text{FIG}\left(a_j, a_b\right) = \text{FE}\left(a_j\right) - \text{FCE}\left(a_j, a_b\right) \qquad (13.3)$$

where $\text{FIG}(a_j, a_b)$ is the functional information gain, $\text{FE}(a_j)$ is functional entropy of the attribute a_j, $\text{FCE}(a_j, a_b)$ is the functional conditional entropy of a_j and a_b.

$$\text{FCE}\left(a_j, a_b\right) = \sum_{j=1}^{u(a_j)} \text{prob}\left(a_j = j, a_b = j\right)\text{cf}\left(\text{prob}\left(a_j = j, a_b = j\right)\right) \qquad (13.4)$$

$\text{FCE}(a_j, a_b)$ is the functional conditional entropy of a_j and a_b, cf is the functional tangent conditional entropy.

$$\text{cf}\left(\text{prob}\left(a_j = j, a_b = j\right)\right)$$
$$= \frac{1}{2}\left[\log\left[\text{prob}\left(a_j = j, a_b = j\right)\right] + \frac{-1}{a\tanh\left(\text{prob}\left(a_j = j, a_b = j\right)\right)}\right] \qquad (13.5)$$

cf is the functional tangent conditional entropy, $\log(\cdot)$ is the logarithmic function, and $a\tanh(\cdot)$ is the inverse hyperbolic tangent function.

Block Diagram
See Fig. 13.2.

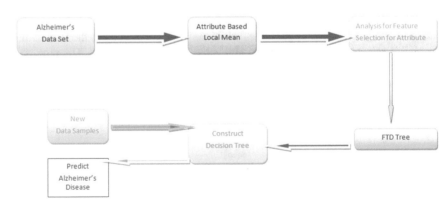

Fig. 13.2 Block diagram of FTD tree

Detailed Study of Alzheimer's Data
Building Tree
Now, before any of the nascent tree structures were created, Alzheimer's data at the root comprised ten yes's and seven no's corresponding to an functional entropy [10,7] = 1.9592 bits.

We calculate the average information value of these, talking into account the number of instance that go down each branch (Tables 13.2 and 13.3; Figs. 13.3 and 13.4).

Statistical Analysis of Alzheimer's Disease
See Figs. 13.5, 13.6, 13.7 and 13.8.

Performance Measure Alzheimer's Disease
Accuracy
According to the accuracy, it is the ability of Alzheimer's to correct and predict the disease of new or previous data (Table 13.4).

$$TP = (TP + TN)/(TP + TN + FP + FN) \qquad (13.6)$$

Precision
Precision is a fraction of relative Alzheimer's disease records among the fetching disease records (Fig. 13.9).

$$PrecisionTP = TP/(TP + FP) \qquad (13.7)$$

Recall
However, recall is the fraction of Alzheimer's disease records that have been fetched over with total amount of admissible disease data.

$$Recall\ (TPR) = TP/(TP + FN) \qquad (13.8)$$

F-measure
F-measure is a weighted harmonic mean of precision and recall of Alzheimer's disease.

$$F\text{-measure} = 2 * (precision * recall)/(precision + recall) \qquad (13.9)$$

Error-rate
In this, the misclassification rate will be considered as Alzheimer's disease data (Fig. 13.10).

$$R = 1 - Acc\ (M) \qquad (13.10)$$

where R is the error rate, and Acc (M) is the zccuracy of the selected variables.

Table 13.2 Split based on variables

Split based on variable 'A'	AD = Y	AD = N	Total
<65	4	3	7
65–85	3	2	5
>85	3	2	5

Split based on variable 'CG'	AD = Y	AD = N	Total
Yes	8	0	8
No	2	7	9

Split based on variable 'IB'	AD = Y	AD = N	Total
Yes	8	2	10
No	2	5	7

Split based on variable 'MD'	AD = Y	AD = N	Total
Yes	7	3	10
No	3	4	7

Split based on variable 'G'	AD = Y	AD = N	Total
M	1	6	7
f	9	1	10

Table 13.3 Functional Information Gain (FIG) of decision variables

Age
FE $[3,4] = 1.9633$ bits
FE $[2,3] = 1.9559$ bits
FE $[2,3] = 1.9559$ bits
$FCE(A) = 7/17(1.9633) + 5/17(1.9559) + 5/17(1.9559)$
   $= 1.9588$ bits
$FIG(A) = FE(AD) - FCE(A)$
   $= 1.9592 - 1.9588$
   $= 0.0004$ bits

IB
FE $[8,2] = 1.8233$ bits
FE $[2,5] = 1.8993$ bits
$FCE(IB) = 10/17(1.8233) + 7/17(1.8993)$
   $= 1.8545$ bits
$FIG(IB) = FE(AD) - FCE(IB)$
   $= 1.9592 - 1.8545$
   $= 0.1047$ bits

G
FE $[1,6] = 1.7505$ bits
FE $[9,1] = 1.6785$ bits
$FCE(G) = 7/17(1.7505) + 10/17(1.6785)$
   $= 1.708$ bits
$FIG(G) = FE(AD) - FCE(G)$
   $= 1.9592 - 1.708$
   $= 0.2512$ bits

CG
FE $[8,0] = 0$ bits
FE $[2,7] = 1.8464$ bits
$FCE(CG) = 8/17(0) + 9/17(1.8464)$
   $= 0.9775$ bits
$FIG(CG) = FE(AD) - FCE(CG)$
   $= 1.9592 - 0.9775$
   $= 0.9817$ bits

MD
FE $[7,3] = 1.9089$ bits
FE $[3,4] = 1.9633$ bits
$FCE(MD) = 10/17(1.9089) + 7/17(1.9633)$
   $= 1.9312$ bits
$FIG(MD) = FE(AD) - FCE(MD)$
   $= 1.9592 - 1.9312$
   $= 0.028$ bits

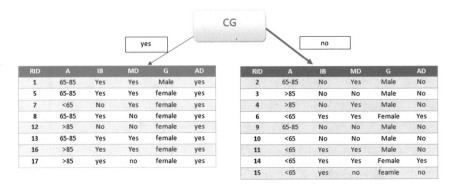

RID	A	IB	MD	G	AD
1	65-85	Yes	Yes	Male	yes
5	65-85	Yes	Yes	female	yes
7	<65	No	Yes	female	yes
8	65-85	Yes	No	female	yes
12	>85	No	No	female	yes
13	65-85	Yes	Yes	female	yes
16	>85	Yes	Yes	female	yes
17	>85	yes	no	female	yes

RID	A	IB	MD	G	AD
2	65-85	No	Yes	Male	No
3	>85	No	No	Male	No
4	>85	No	Yes	Male	No
6	<65	Yes	Yes	Female	Yes
9	65-85	No	No	Male	No
10	<65	No	No	Male	No
11	<65	Yes	Yes	Male	No
14	<65	Yes	Yes	Female	Yes
15	<65	yes	no	feamle	no

Fig. 13.3 Splitting on the attribute that provides the highest functional information gain

Fig. 13.4 Alzheimer's disease decision tree based the training set

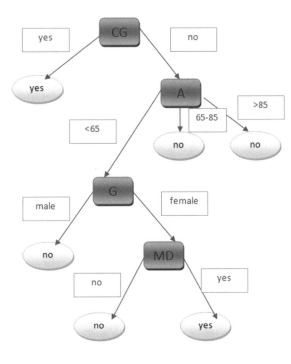

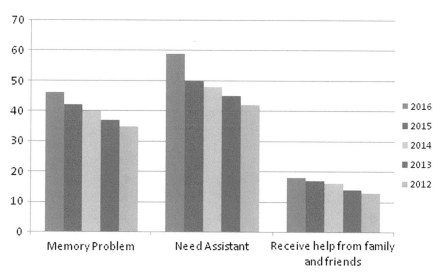

Fig. 13.5 Memory problems who say about certain difficulties

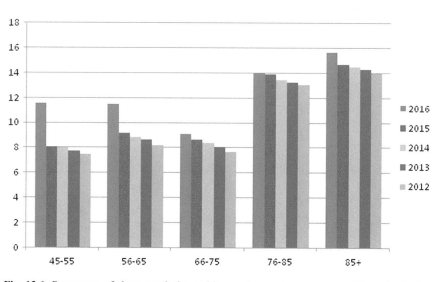

Fig. 13.6 Percentage of those aged above 45+ people worries memory problems in Andhra Pradesh

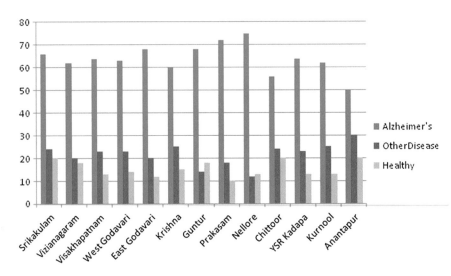

Fig. 13.7 Suffering from Alzheimer's disease in Andhra Pradesh different district compare to other disease, healthy

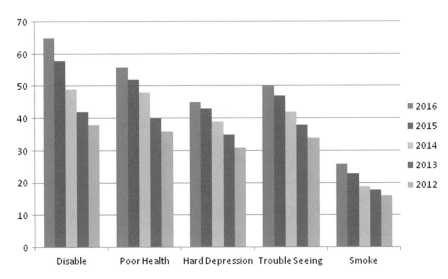

Fig. 13.8 Characteristics of those with memory problems in Andhra Pradesh

Table 13.4 Alzheimer's accuracy measures

Models	ID3	C4.5/J48	CART	FTD
Accuracy	82.35	76.47	82.74	90.1
Precision	0.833	0.667	0.7	0.9
Recall	0.7142	0.857	1	0.89
F-measure	0.769	0.75	0.824	0.9
Error-rate	0.1765	0.2353	0.2353	0.09

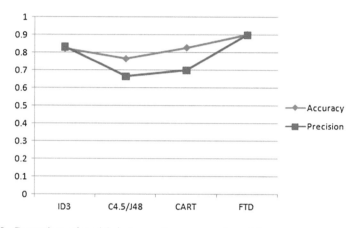

Fig. 13.9 Comparison of models in terms of accuracy and precision

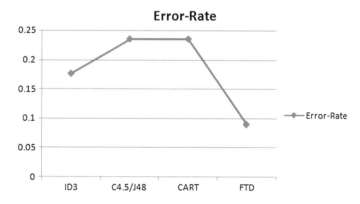

Fig. 13.10 Comparison in terms of error rate

13.6   Conclusion

The uncertain data is used to predict the Alzheimer's disease from the Alzheimer's dataset. After the identification of important features, the classification is performed. The FTD Tree introduced two new subroutine called functional tangent entropy and splitting rules. Once the FTD Tree is constructed for predicting the disease, the physicochemical parameters are analyzed. For scientific research, the five physicochemical parameters are taken from ponds. The performance of the overall proposed system is examined with the existing classifier algorithm with a help of the parameters, such as sensitivity, specificity, and accuracy. The FTD Tree system reaches the maximum classification accuracy of 95% as compared with the existent work. To obtain optimization algorithm for predicting the Alzheimer's disease in future, the best split point can be found out. The best split point for the FTD Tree algorithm may improve the prediction quality.

References

1. National Center for Biotechnology Information. http://www.ncbi.nlm.nih.gov
2. Quinlan JR (1986) Induction on decision tree. Mach Learn 1:81–106
3. Quinlan JR (1993) C4.5: programs for machine learning. Morgan Kaufmann
4. Kadaru BB, Uma Maheswara Rao M, Narni SC (2017) AFI-ID3 data mining algorithm for conducting cardiac test. IJCMS 6(9)
5. Kadaru BB, Narni SC, Raja Srinivasa Reddy B (2017) Classifying Parkinson's disease using mixed weighted mean. IJAER 12(1)

Chapter 14
Experimental Investigation of Cognitive Impact of Yoga Meditation on Physical and Mental Health Parameters Using Electro Encephalogram

Ramasamy Mariappan and M Rama Subramanian

Abstract Nowadays, our human life becomes machinery and increases mental stress. Yoga Meditation relieves our mental stress and re-establishes mental harmony when practiced regularly, and hence, it promotes physical, mental, and spiritual health. The previous studies on meditation show that there is relationship between mindful meditation, brain, and heart. As the meditation has direct impact on the mind power, the mental health is influenced by the depth of meditation. Also, brain and mind has the direct control over the heart, and physical health is indirectly related by mindful meditation. Experimental investigations on the effect of meditation are proposed by monitoring electroelectroencephalogram (EEG) with using brain–computer interface (BCI). The brain waves of the person who is under meditation will be captured using Neurosky BCI mind wave kit. The mental health status is analyzed by monitoring alpha, beta, and theta brain waves during long meditation. The impact of mediation on the heart rate variation (HRV), respiration rate, and blood pressure, etc., is monitored to evaluate the effectiveness of meditation on the physical health status. The captured brain waves through Neurosky BCI mindwave kit are filtered to get alpha and beta waves. The alpha and beta brain waves are analyzed using BCI Graphical User Interface (GUI) software tool to investigate the power levels of each wave and duration before and after meditation. The observed data was further analyzed through IBM Statistical Package for the Social Sciences (SPSS) tool for studying about mean, standard deviation, and variance of brain wave parameters, which shows the positive impact on physical as well as mental health parameters, especially cardiovascular (CV), and heart rate variation (HRV) parameters, etc.

R. Mariappan (✉)
Research and Development Centre, Sri Venkateswara College
of Engineering, Tirupati, Andhra Pradesh, India
e-mail: mrmrama2004@rediffmail.com

M Rama Subramanian
Department of Electronics and Communication Engineering,
Madras Institute of Technology, Chennai, Tamilnadu, India
e-mail: rama.mit2016@gmail.com

© The Author(s) 2019
N. B. Muppalaneni et al., *Soft Computing and Medical Bioinformatics*,
SpringerBriefs in Forensic and Medical Bioinformatics,
https://doi.org/10.1007/978-981-13-0059-2_14

Keywords   Meditation · Yoga · Electroelectroencephalogram (EEG)

14.1   Introduction

Yoga meditation is the science of peace in one's personal and social life, which helps to maintain stability of mind in the adverse circumstances and to attain the highest level of emotional equilibrium. Yoga Meditation gives us mental relaxation as well as physical relaxation and constant peace and improves the quality of human being. It is the detergent of mind and purifies our mind, and it makes us to experience our true mind power and real and lasting peace with great bliss. Mediation Yoga spreads vibrations of peace, transforming into peaceful humans with positive thoughts. Yoga is concerned with focusing the mind and using its power to control the body. There should be a relationship between the mind and body, which in turn relates mental health and physical health. Through the Yoga Meditation, an individual will get relief from mental depression, tension, negative thoughts, sins, bad habits, freedom from addictions, etc. Thus, proper and regular Yoga Meditation practice will yield good physical health including good heart rate, normal blood pressure.

Scientific studies have investigated that meditation reduces blood pressure as well as symptoms of ulcerative colitis. Recent researches on brain wave monitoring have proven that long-term meditations would profound impact on the brain wave activity. It also investigated that the activity of alpha–theta is predominant wave during meditation. In recent years, the use of modern scientific techniques and instruments, such as functional Magnetic Resonance Imaging (fMRI) and EEG which are able to directly observe brain functionality and neural activity during and after meditation. Exploratory brain analyses investigated that significant increase in gray matter concentration in the brain. From these studies, it is evident that there is a relationship between meditation, brain and heart and physical activities. Hence, there is a need to investigate the impact of Yoga meditation on mind power through brain wave analysis and hence to establish the relation between meditation, heart and physical as well as mental health.

The outline of the paper is as follows. Section 14.2 reviews the current literature in the field; Sect. 14.3 defines the problem and its solution, followed by the implementation methodology and experimental design. Section 14.4 shows the results and discussion, followed by conclusion in Sect. 14.5.

14.2 Related Work

This section summarizes the brief review of literature in the field of Meditation. There are many studies in current literature [1–5] which studied the benefits of yoga meditation in India as well as abroad. Each study focused on impact of Yoga [6, 7] on reduction of stress, improvement of physical health, mental health, etc. Some studies revealed that the Meditation improved the cognition ability in terms of attention and meditation levels [8, 9]. Few researchers have studied about the mind–hand coordination and mind–heart coordination with the help of Meditation. However, there are very few studies only attempted about the impact of Raja Yoga Meditation in India and abroad [10–15, 16].

Recent studies show on Raja Yoga meditation increase of heart rate and blood pressure. The regular practicing Raja Yoga meditation is a simple way to balance our physical state, cognitive state, and emotional state in our daily life. The Raja Yoga meditation [17] can serve as an auxiliary tool to the conventional medicine for the physical and mental health and also as a preventive tool. We should replace our mechanistic approach to study of health and disease with the humanistic one.

Numerous researchers [18–20] have assessed the efficacy of yoga meditation as an adjunct to routine management of various diseases and disorders. A new research study shows that a little yoga meditation everyday keeps the doctor away. Madanmohan et al. have reported the effect of yoga Meditation prevents diabetic mellitus. He found a significant decrease in the fasting and post-prandial glucose level. Many scientific reports have proven that *"**Brainwaves affect Mental States**"* and also, *"**Mental states affect Brainwaves**."* The major benefits of yoga Meditation are as follows.

- Meditation improves heart health and helps to relieve depression and anxiety.
- Meditation reduces the metabolic rate, heart rate and reduces heart disease.
- Yoga Meditation reduces stress by lower levels of stress-hormones.
- Meditation reduces high blood pressure and enhances energy, strength, and fitness.
- Yoga Meditation helps asthma patients, making breathing easier.

There is a huge need to support the above benefits of meditation through scientific investigations of physical and mental health parameters. Hence, there is a need for scientific techniques, innovative methodologies, and interventions for investigating the effect of Yoga meditation. This research work contributes to investigate the impact of Raja Yoga on physical and mental health parameters using EEG-based brain–computer interface (BCI) by monitoring brain waves during and after meditation. There are several studies which investigated the effect of yoga/meditation on physical and mental health as summarized in Table 14.1.

Table 14.1 Literature review of national research papers

Ref. ID	Author	Year	Methods	Problems found
1.	Patl Girish	1984	Clinical measurements of cardiovascular parameters	No experimental investigations
2.	Shirley Telles and T. Desiraju	1993	Autonomic and respiratory variables	No study on EEG parameters
3.	Patel Nikhil	1996	Neurotic illnesses and changes in physiological parameters	No specific study on EEG, ECG, etc.
4.	Surya, Balakishore	2000	TAT test and symptom list	Statistical test only done
5.	Sukhsohale Neelam D. and Phatak Mrunal S.	2001	Physiological and psychological effects	No experimental investigations
6.	Kwanchanok Boonjaksilp, Vichit Punyahotra	2007	Brainwave activities and stress reduction	Clinical studies only done
7.	Jayashree Santhosh, Gracee Agrawal, Manvir Bhatia, Nandeeshwara S. B., Sneh Anand	2016	Temporal EEG spectral analysis	EEG analysis and physical health parameters not considered

14.3 Proposed Work

As the brain is center controller for functioning of the all the organs of the human body, yoga Meditation has positive impact on the mental health as well as physical body. The existing literature shows that many researchers studied the beneficial efficiency of meditation on mental disorders and physical diseases, in addition to the routing management of diseases. The proposed research study is focused at investigating the impact of Raja Yoga Meditation on mental health parameters and physical health parameters namely respiratory functions, cardiovascular parameters, and lipid profile. The electrical activity of the brain can be mapped using variation of electroencephalogram (EEG) signal pattern during meditation. Thus, we can analyze the effects of meditation on mental health of a person using EEG signals, and hence, there is a debate on the EEG brain wave pattern changes during meditation and signal processing challenges, etc.

The current literature studied only particular type of Meditation and not studied about other forms of meditation. As there is a strong connection between brain, mind, and heart, this research work first establishes the scientific relationship between Yoga meditation, mental health, and physical health of well-being with using brain–computer interface (BCI) model and associated tools to measure the effectiveness of Yoga meditation. This research work proposes to experimental investigation of the impact of Raja Yoga meditation on physical and mental health over short-term as well as long-term meditation using EEG signal analysis of brain wave variation with BCI and cardiovascular parameters such as heart rate variation (HRV), blood pressure with appropriate sensors. The major objectives of the proposed research work are listed as follows.

Objectives

i. To investigate the impact of Raja Yoga meditation on physical and mental health by monitoring EEG signals using brain–computer interface (BCI) and its associated tools.

ii. To experimentally investigate the relationship between meditation, mind, and heart and hence the relationship between meditation, mental health and physical health.

iii. To investigate the impact of Raja Yoga meditation on mental health using BCI by monitoring alpha and beta brain wave signals of EEG.

iv. To measure the cardiovascular parameters such as heart rate, heart rate variation, blood pressure during and after the Yoga Meditation using bio-sensors.

v. To analyze the brain wave signals and cardiovascular parameters in terms of their mean, standard deviation, variance, etc.

Hypothesis

There is significant effect of Raja Yoga meditation on mental and physical health of humans.

14.3.1 Implementation Methodology

The investigation on the impact of Raja Yoga on physical and mental health will be carried out on a two group of people, one group with short-term meditation with 1 or 2 years of experience and another group with long-term meditation with more than 10 years. This research project proposes to investigate the impact of Raja Yoga meditation on physical and mental health over short-term as well as long-term meditation using EEG signal analysis of brain waves with BCI and cardiovascular parameters such as heart rate variation (HRV), blood pressure with sensors. The brain wave analysis can be done using EEG signal analysis with the help of Neurosky brain wave kit. The detailed methodology for this research study is described below.

14.3.1.1 Mental Health Parameters—Brain Waves Using Neurosky BCI

The Neurosky brain–computer interface used for capturing the alpha, beta, and theta waves is done by using **Neurosky BCI mindwave kit**, which is wearable embedded sensor device like regular headphones. It has a microchip which preprocesses the microvolt-level brain wave signal and transmits that EEG data to the computer interface through Bluetooth.

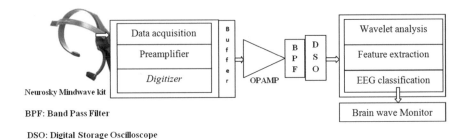

Fig. 14.1 Brain wave monitoring using Neurosky mindwave kit

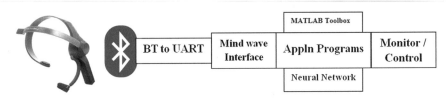

Fig. 14.2 Brain wave monitoring using MATLAB

14.3.1.2 EEG Data Acquisition and Preprocessing

As shown in Fig. 14.1, the captured EEG signal is preprocessed by pre-amplifier and digitizer and then passed through a buffer followed by an operational amplifier (OPAMAP). The amplified EEG signal is then passed through a band pass filter (BPF), which passes the band of frequencies between 0.5 and 40 Hz and also removes the 50 Hz power line signal present in the EEG signal. The amplified and filtered EEG signal is fed into Digital Storage Oscilloscope (DSO) and then applied signal processing for feature extraction and classification of brain waves as shown in Fig. 14.1. The brain wave is monitored using Neurosky mindwave interface with application programs built with MATLAB toolbox and neural network as shown in Fig. 14.2.

14.3.1.3 EEG Signal Analysis and Feature Extraction

After passing through noise filtering and amplification, the EEG signal is applied through further steps for signal processing. As the brain waves fall in different frequency bands, time–frequency analysis helps in characterizing EEG signal analysis with good time as well as frequency resolution. The algorithm for analyzing brain waves with MATLAB using wavelet transform is shown in Fig. 14.3.

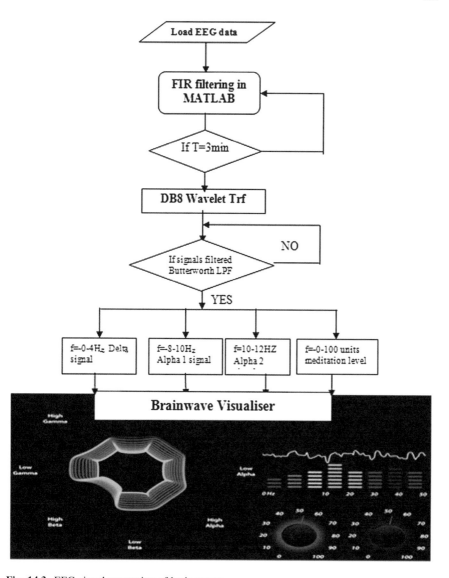

Fig. 14.3 EEG signal processing of brain waves

14.4 Results and Discussion

Using Neurosky BCI kit, digital EEG and its associated tools are used to analyse the brain wave signals in different ways based on the parameters need to be studied or analyzed as shown in Fig. 14.4. The various parameters considered are listed below.

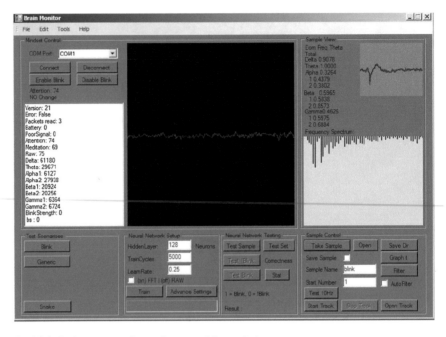

Fig. 14.4 Brain wave monitor software tool for analysis

14.4.1   Performance Measures

The following parameters are considered for studying heart rate variation (HRV) in time domain, frequency domain, and time–frequency domain as shown in Fig. 14.5.

Time Domain HRV

(1) **mRR**: Mean of interbeat intervals,
(2) **mHR**: Mean of heart rates,
(3) **sdNN**: Standard deviation of interbeat intervals,
(4) **sdHR**: Standard deviation of heart rates,
(5) **RMSSD**: RMS sum of the squares of adjacent interbeat interval differences.

Frequency Domain HRV

(1) **VLF**: Power spectrum in VLF range,
(2) **LF**: Power spectrum in LF range,
(3) **HF**: Power spectrum in HF range,
(4) **LF/HF**: Ratio between power spectrum in LF and n HF ranges.

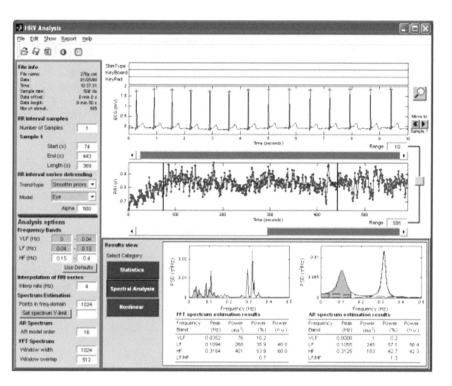

Fig. 14.5 Heart rate variability analysis during meditation

Time–Frequency Domain Parameters

(1) Width of the highest spectral peak (**PeakW**),
(2) Frequency of the highest spectral peak (**PeakF**),
(3) Minimum, maximum heart rates,
(4) Mean heart rate variation (HRV),
(5) Blood pressure, systolic blood pressure, diastolic blood pressure, etc.

14.4.2   Statistical Analysis

The heart rate variability test was carried out during meditation and the EEG signal analysis was shown in Fig. 14.5. It shows that the mean value of HRV increases with meditation time. The oxyhemoglobin as well as total oxyhemoglobin level of blood increases with different states of Meditation as shown in Fig. 14.6. The statistical analysis of Meditation and EEG data along with health parameters was done using SPSS. The pre- and post-Meditation values of HRV were tabulated, and coefficient of variation was calculated to evaluate pattern of distribution. Mean Standard deviation, τ^2 test, Chi^2 were analyzed. Performance of both the

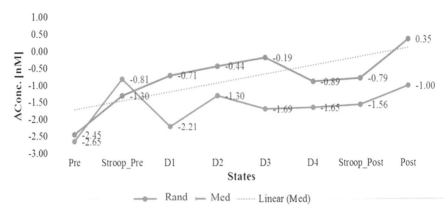

Fig. 14.6 Oxyhemoglobin for different states of meditation

groups on study variables is compared using the independent sample t-test. Finally, NIMANS Neuropsychological Battery were used to assess cognitive functions on physical as well as mental health parameters.

14.5 Conclusion

This research paper proposed a new method of investigating the cognitive impact of yoga Meditation on physical and mental health of human. The brain wave consists of alpha, beta, gamma, theta, and delta, among which alpha and theta waves are more predominant during yoga meditation. The best tool for acquiring these brain waves is electroencephalogram (EEG) which was captured by Neurosky brain wave kit, which in turn processed by brain wave monitoring tool, e-Sense, and its associated tools. The experimental investigations through BCI-based EEG show that there is a huge cognitive impact of yoga Meditation on physical and mental health parameters, which are analytically, studied through respective parameters namely blood pressure, cardiovascular, and heart rate variability (HRV) parameters.

References

1. Stam CJ (2014) Modern network science of neurological disorders. Nat Rev Neurosci 15:683–695
2. Lou HC, Kjaer TW, Friberg L, Wildschiodtz G, Holm S (1999) 150-H2O PET study of meditation and the resting state of normal consciousness. Hum Brain Map 98–105
3. The Venerable Mahasi Sayadaw (1965) The progress of insight (Visuddhinana-katha)
4. Kasamatsu A, Hirai T (2015) An electroencephalographic study of Zen meditation (zazen). Psychologia 12.225, 143

5. Scientist (1978) Effect of Rajyoga meditation on brain waves (EEG). Langley Porter Psychiatric Institute, San Francisco, California, and USA
6. Madhavi Kanakdurga G, Vasanta Kumari Dr. (2012) Attention regulation of mediators and non-mediators of class IX. Indian J App Resh 1(5)
7. Brown DP, Engler J (1980) The stages of mindfulness meditation: a validation study. J Transpers Psychol 12:143–192
8. Kabat-Zinn J (1982) An outpatient program in behavioral medicine for chronic pain patients based on the practice of mindfulness meditation: theoretical considerations and preliminary results. Gen Hosp Psychiatry 4:33–47
9. Shapiro DH (1982) Overview: clinical and physiological comparison of temporal uncertainty and filter strategies. Am J Psychiatry 139(3):267–274
10. Cranson R, Goddard PH, Orme-Johnson D (1990) P300 under conditions of temporal uncertainty and filter attenuation: reduced latency in long-term practitioners of TM. Psychophysiology 27:S23
11. Telles S, Joseph C, Venkatesh S, Desiraju T (1992) Alteration of auditory middle latency evoked potentials during yogic consciously regulated breathing and attentive state of mind. Int J Psychophysiol 14:189–198
12. Telles S, Desiraju T (1993) Autonomic changes in Brahmakumaris raja yoga meditation. Int J Psychophysiol 15:147–152
13. Valentine ER, Sweet PLG (1999) Meditation and attention: a comparison of the effect of concentrative and meditation on sustained attention. Ment Relig Cult 2
14. Buchheld N, Grossman P, Walach H (2001) Measuring mindfulness in insight meditation and meditation based psychotherapy: the development of the Freiburg Mindfulness Inventory (FMI). J Medit Medit Res 1
15. Ferrarelli F, Smith R, Dentico D, Riedner BA, Zenning C, Benca RM, Lutz A, Davidson RJ, Tononi G (2013) Experienced mindfulness exhibit higher parietal-occipital EEG gamma activity during NREM sleep. Open Access PLOS 8(8)
16. Aftanas LI, Goloshekin SA (2003) Changes in cortical activity in altered states of consciousness: the study of meditation by high resolution EEG. Hum Physiol 28(2):143–151
17. Baer RA, Smith GT, Allen KB (2004) Assessment of mindfulness by self report: the Kentucky inventory of mindfulness skills. Assessment 11:191–206
18. Babiloni C, Carducci F, Lizio R et al (2013) Resting state cortical electroencephalographic rhythms are related to gray matter volume in subjects with mild cognitive impairment and Alzheimer's disease. Hum Brain Mapp 34(6):1427–1446
19. Ivanovshi B, Malhi GS (2007) The psychological and neurophysiological concomitants of mindfulness forms of meditation. Acta Neuropsychiatr 19:76–91
20. Jayashree Santhosh A, Gracee Agrawal, Manvir Bhatia, Nandeeshwara SB, Sneh Anand (2016) Extraction and analysis of EEG waves for the study of enhancement of brainwaves through meditation. IJPT, pp 13481–13488

Printed in the United States
By Bookmasters